MODERN EUROPEAN PHILOSOPHY

BACHELARD: SCIENCE AND OBJECTIVITY

BACHELARD: SCIENCE AND OBJECTIVITY

MARY TILES

The right of the
University of Cambridge
to print and sell
all manner of books
was granted by
Henry VIII in 1534.
The University has printed
and published continuously
since 1584.

CAMBRIDGE UNIVERSITY PRESS

CAMBRIDGE
LONDON NEW YORK NEW ROCHELLE
MELBOURNE SYDNEY

CAMBRIDGE UNIVERSITY PRESS
Cambridge, New York, Melbourne, Madrid, Cape Town, Singapore, São Paulo

Cambridge University Press
The Edinburgh Building, Cambridge CB2 2RU, UK

Published in the United States of America by Cambridge University Press, New York

www.cambridge.org
Information on this title: www.cambridge.org/9780521289733

First published 1984

A catalogue record for this publication is available from the British Library

Library of Congress Catalogue Card Number: 84–5001

ISBN-13 978-0-521-28973-3 paperback
ISBN-10 0-521-28973-4 paperback

Transferred to digital printing 2006

For John Tollyfield
my father

CONTENTS

EDITORS' INTRODUCTION

The purpose of this series is to help to make contemporary European philosophy intelligible to a wider audience in the English-speaking world, and to suggest its interest and importance, in particular to those trained in analytical philosophy. The first book in the series was, appropriately enough, Charles Taylor's book *Hegel and Modern Society*. It is by reference to Hegel that one may indicate most starkly the difference between the two traditions to whose intercommunication the series seeks to contribute; for the analytical philosophy of the contemporary English-speaking world was largely developed by Moore, Russell, and others in revolt against idealism and the dominance of British Hegelians at the turn of this century. It was not a natural development of a well established empiricist tradition, as some have claimed. It is true that the British and American idealists had themselves already diverged considerably from Hegel, but their holistic philosophy was certainly Hegelian both in terminology and in aspiration. Moore and Russell, for their part, also owed much to a different tradition, one stemming from Hume. Nevertheless, it should not be forgotten that they too were influenced by European contemporaries, to whose writings, indeed, they explicitly appealed in their revolt against the British Hegelians. In particular they admired two European philosophers who had very little sympathy for Hegelianism: Brentano in the case of Moore and, in the case of Russell, Frege.

The next book in our series, Raymond Geuss's *The Idea of a Critical Theory*, discussed issues which could be understood only by reference back to the thought of Hegel and Marx, which – in marked contrast to what happened in the English-speaking world – was absorbed and further developed in various ways, by the main trends of radical thought on the European continent. In England, on the other hand, philosophical opposition to 'Establishment' ways of thinking and patterns of influence was developed in opposition to Hegel rather than under his influence. In the mid-thirties, just when a new Hegelian social philosophy was being developed in Frankfurt, and Hegel's philosophy was being introduced seriously to the academic world in Paris, A. J. Ayer brought back from Vienna to Oxford logical positivism one of whose chief targets of attack was Hegel. It is true that logical positivism was short-lived in England; and even in the United States, to which distinguished members of the Vienna Circle eventually escaped, it represented an important phase rather than a lasting school. But many of the philosophical virtues with which it was most concerned continued to be fostered. What is now called analytical philosophy, with its demand for thoroughness of conceptual analysis and its suspicion of rhetoric and grandoise structures, came to be more and more dominant in the English-speaking world. The philosophical attitude which it represents and which distinguishes it from the dominant European schools of thought is succinctly expressed in the foreword to the *Philosophical Remarks* (1930) of Wittgenstein, whose influence on analytical philosophy was incalculable:

> This spirit is different from the one which informs the vast stream of European and American civilisation in which all of us stand. The one spirit expresses itself in an onwards movement, in building ever larger and more complicated structures: the other in striving after clarity and perspicuity in no matter what structure... And so the first adds one construction to another, moving on and up, as it were, from one stage to the next, while the other remains where it is and what it tries to grasp is always the same.

There are several strands of recent Continental European philosophy that have remained unnoticed by mainstream analytic philosophers, although they in fact concern themselves with problems that have been central to the analytic tradition. The next two works in our series (Lambert and Tragesser) as well as this book by Mary Tiles bring these strands to our attention. Professor Lambert showed the relevance of the Principle of Independence (that is, the independence of the *Sosein*, or nature of an object from its being) to theories of predication and quantification. Professor Tragesser brings the work of Husserl to bear on issues in the philosophy of mathematics. Up to now Husserl's work has had some influence on analytic philosophy in the general areas of philosophy of mind and value-theory, but Professor Tragesser demonstrates the relevance of Husserl's views to such issues as the problem of reference, the relation of truth conditions to assertability conditions, and questions of objectivity and realism.

The late Gaston Bachelard's lectures in the Sorbonne as well as his many written works covered a very wide field from the conceptual implications of quantum physics to poetry. The reception of his thoughts could not have been predicted. On the one hand, although Bachelard was no Marxist, various of his views were appropriated by a whole generation of Marxists. On the other hand, the impact he made on analytical philosophy in the central area of his interest, philosophy of science was, quite unjustly, negligible. His historicism which preceded that of Kuhn or Foucault has never been properly discussed. Some people may have heard of his notion of 'epistemological rupture' through the works of Louis Althusser, but without giving serious attention to it – or even associating it with Bachelard himself.

Mary Tiles shows in this book how Bachelard's views are related to the concerns which analytic philosophers have about the status of science and rationality, and their debates concerning realism, operationalism and relativism. Dr Tiles says that hers is not an attempt to reconstruct Bachelard's position in the language of analytical philosophy, since Bachelard makes one question many of the presuppositions behind analytical philosophy itself. She shows, nevertheless, that Bachelard's position is a natural development of a view of Duhem, who has had considerable influence on analytical philosophy of science, according to which to give the history of physical principles is at the same time to make a logical analysis of them.

According to Bachelard, the rationality of science is not so much to be judged in terms of its content as by the norms of its thought and practice; and this is revealed by the way in which its concepts emerge from a sequence of corrections. Bachelard's doubts about 'thing' or 'substance' oriented ways of thinking ('*chosisme*'), his rejection of foundationalism or instrumentalism as well as of views of rationality that make it entirely culture relative, are all close to the issues being reexamined by analytical philosophers now. Dr Tiles makes this clear by relating many of Bachelard's arguments to those of thinkers like Putnam, Lakatos, Feyerabend or Van Fraassen. Bachelard's claim that commitment to rationality does not remove the element of choice or voluntariness in the development of scientific thinking is of special interest to present day discussions.

The labels 'analytical' and 'continental' are in many ways very unsatisfactory. There are philosophers of the phenomenological tradition working in the United States. There are many other philosophers engaged in work of conceptual analysis in the Scandinavian countries, Poland, and (more recently) Germany. Moreover, the universities of Europe which have not been influenced by the analytical tradition – and these include nearly all of those in France and Italy, and the great majority of those in German-speaking countries and in Eastern Europe – have themselves by no means represented any unitary tradition. Yet the disagreements or even lack of communication between, for instance, Hegelians, Marxists, phenomenologists and Thomists are 'small' in comparison with the barriers of mutual ignorance and distrust between the main representatives of the analytical tradition on the one hand and the main philosophical schools of the European continent of the other (schools which are also dominant in Latin America, Japan, and even some universities in the United

States and Canada). These barriers are inevitably reinforced by the fact that, until very recently at any rate, even the best students from the universities situated on either side have tended to emerge from their studies with such divergent areas of knowledge and ignorance, competence and incompetence, that they are hardly equipped even to enter into informed discussion with each other about the nature of what separates them.

We tend, nevertheless, to forget that the erection of these barriers is a relatively recent phenomenon. Brentano, writing on the philosophy of mind at the end of the last century, made frequent reference to J. S. Mill and to other contemporary British philosophers. In turn, as we have noticed, Moore refers to Brentano. Bergson discusses William James frequently in his works. For Husserl one of the most important philosophers was Hume. The thinkers discussed seriously by Russell include not only Frege and Poincaré, but also Meinong. How unfortunate, then, that those who have followed in their footsteps have refused to read or to respect one another, the one group convinced that the other survives on undisciplined rhetoric and an irresponsible lack of rigour, the other suspecting the former of aridity, superficiality, and over-subtle trivialisation.

The books of this series represent contributions by philosophers who have worked in the analytical tradition but who now tackle problems specifically raised by philosophers of the main traditions to be found within contemporary Europe. They are works of philosophical argumentation and of substance rather than merely introductory résumés. We believe that they may contribute towards the formation of a richer and less parochial framework of thinking, a wider frame within which mutual criticism and stimulation will be attempted and where mutual disagreements will at least not be based on ignorance, contempt or distortion.

PREFACE (AND POSTSCRIPT)

It is only fair that the reader should be warned that this book is *not* a work of definitive exegesis; it could not hope to be, given the gulf between the author's philosophical background and Bachelard's. The *Bachelard* who inhabits its pages is a rational construct. Bachelard himself, by producing texts on the philosophy of science, provided the materials for this construction.[1] This was not a free creative act; it was contextually constrained, but not textually determined.

What could be the object of such an exercise? What justification can it be given? It is an attempt to confront the phenomenon which has been labelled the 'incommensurability of scientific theories', the phenomenon which has been perceived as a threat to the possibility of objectivity in science, and to do so at two levels. Most straightforwardly this is carried out by looking at the works of a philosopher of science who would appear to have resolved the problem. For although Bachelard recognises the phenomenon under the guise of discontinuity in the history of science, not only does he not see it as a threat to the possibility of rationality or objectivity in science, he sees it as a product of the rational processes of the advancement of science. But approach to this philosopher could only be made across a gulf similar in kind to that which exists between 'incommensurable' theories. The approach via rational construction is thus itself an exploration of the extent to which, and the means by which, such gulfs may be crossed.

[1] Bachelard also wrote extensively on literature and on poetic expression, and is better known in the English-speaking world for these works. Although I have indicated points at which these apparently divergent concerns make contact, I have not explored this dimension of Bachelard's philosophy, although it would be necessary to do so for any complete exegetical account.

But if this was the project, the course taken in its actual execution raises certain problems for the book's own evaluation, problems which it would have been difficult to acknowledge in the text. In the first place, having sought to (re)construct a position which sees rationally justifiable discontinuities where others see incommensurabilities, it concludes by being unable to find a neutral standpoint from which to evaluate the merits of the (re)constructed position. The practical attempt at philosophic commensuration seems, therefore, to have failed; the book, in effect, appears to pass negative judgement on itself.

However, it is also part of the book's thesis that the possibility of commensuration does not rest on the availability of a neutral standpoint, for no such standpoint is available. So, in the second place, we have to ask 'From what standpoint are the book's conclusions reached?' The book seems to adopt a pose of neutrality to which, if its content is to be believed, it has no right. Not only has it no right to this pose, it is not, in fact, neutral between the positions discussed.

The project was motivated by a feeling that the discussions of incommensurability were locked in problems which were generated more by the framework of discussion, the ground-rules, than by the phenomenon supposedly under discussion. By being prepared to put that framework in question, the project, in its initial conception, had already moved outside it, but not in any determinate direction. The constructive exercise is thus in part also analytic in function, seeking the questions which it would be profitable to raise, seeking the presuppositions behind analytic philosophy of science which need to be questioned if sense is to be made of Bachelard's texts, and working out some of the consequences of dispensing with those presuppositions. This is a situation and a project which can be understood on the basis of the discussion of 'interference' (pp.104–9). In this case it is the interference set up by the co-presence of two very different treatments of a single subject matter. Given the rational structures from which I, by dint of cultural location, had to start (namely those conferred by a British education in mathematics and its foundations, and in philosophy), the project could, for me, only be to try to reconstruct Bachelard's position using the structures already available to me. Now if what is said in the course of Chapter 3 concerning interference is in any way correct, it would have to be admitted (1) that what is in this way constructed is not, and could not be, identical with the original, (2) that the reconstruction cannot by that admission escape all criticism any more than attempts to construct the continuum out of points can escape criticism, (3) that the reconstruction cannot claim anything like uniqueness, there might have been other ways to go about it, and (4) that the project cannot leave (have left) the antecedent structures in place, but will have forced them to be modified and extended, yet in such a way that these alterations are grounded in, or arise out of, the original structures.

In other words, the general (reconstructed) framework attributed to *Bachelard* is one within which the present project makes some sense and which it therefore in this respect endorses. This does not necessarily extend to endorsement of all the details. From the non-neutral standpoint of the book, from *Bachelard*'s point of view, it is clear that the account of the epistemology of contemporary science is to be assessed by reference to that science and its history; such an assessment cannot dispense with accounts of particular sciences through particular stages of their development. In other words, the account is to be assessed by reference to its subject matter, the phenomena which it seeks to understand. The philosophy of science is not seen as separable from science itself; it belongs with the critical–reflective part of the epistemological process. It is in terms of its ability to yield an understanding of contemporary science in the light of its history, and thus in its historical context, in a way which makes critical evaluation of current theoretical and experimental practices possible that *Bachelard*'s account of science is to be evaluated.

Here there is a respect in which (because of the space available) I have been able to do less than justice to Bachelard's case, and have given a distorted picture of his work. The richness of examples, so characteristic of his works, is lacking in mine. He does not, in any single work, set out to argue systematically for a position. His case on specific points is most frequently made by considering examples. The problem which this presents to one approaching his work from outside the tradition within which he is working is that of knowing quite what to make of these examples and their (usually fairly sketchy) treatment. As is seen (pp.223–5) in the case of the question of the involvement of mathematics in science, the treatment of examples and the conclusion drawn from them are conditioned by the philosophic tradition (the problematic) within which the discussion is being conducted. It is for this reason that I found it necessary to proceed by trying to (re)construct a general position, a framework for discussion. Once we understand this position, we can see that its assessment requires us to go back to examples, to go back to science itself, but this time with specific questions.[2]

But what, then, of the apparent failure of the attempt at philosophic commensuration? The conclusion, in posing the problem of evaluation in the way that it does, is (from its non-neutral standpoint) recognising the difference between construction and argument. Construction is the necessary first, framework-extending step. It is more in the nature of a thought experiment whose function is not just to assist the construction of a new theoretical position, but is also to assist in understanding the nature of that position by setting it in relation to the position from which it started, and hence locating the ways in which the new position may be argued for and tested (see pp.174–9). It does not itself constitute a test or an argument, but serves to suggest where critical attention should be located in order to resolve the previously vague and not well-articulated feeling that something in the established framework, or way of looking at things, needed to be changed (cf. Kuhn 1964). What the experiment reveals is that in order to argue the case for a reappraisal of the distinction between

[2] One has already examples of work in the history and philosophy of science which has been influenced by Bachelard. Most notable (from the point of view of English speakers) is perhaps the work of Alexandre Koyré, who has, in his turn, been a source of inspiration to many others, including Kuhn.

context of justification and context of discovery and of the way in which philosophers approach the history of science (see Chapter 1) it is necessary to reopen discussions of rationality, of the rational subject, and of the role of the subject in the acquisition of objective knowledge. To engage with the framework for analytic philosophy of science we have to take it back to a *critical* discussion of Frege. Frege's achievements in formal logic were monumental, but this does not preclude them from critical evaluation. We need at least to understand their impact, to be aware of the choices made in using this logic to shape the framework of philosophic discussions. The conclusion is not that this is impossible, but that the argumentative work remains to be done.

But it also has to be recognised that a condition of the possibility of making any such argument effective is that there already be openness to at least the possibility of change (p.64). The only possible strategy for arguing that case is by showing that by the standards defining the accepted framework itself there is both the possibility of and the need for change. The problematic character of the particular issue of the relation between philosophy and science is that it is one concerning the nature of philosophy itself, and in particular one which turns as much on the relation between philosophy and conceptions of the human subject as on its relation to science. Discussion of such issues thus encounters all the problems of circularity and reflexiveness. If reflexivity is regarded as incompatible with rationality or the possibility of rational debate, then the framework for such discussion must equally be regarded as unavailable. But that admission can be made only at the cost of a hiatus in the conception of the human subject as also a rational subject. The reflexive character of much of our own thought cannot be denied; philosophers owe their livelihood to it. Are we then doomed to irrationality, to the impossibility of ever making sense of ourselves? What is clear is that the groundrules for discussions of rationality are not clear. The present discussion will have served its purpose if it demonstrates not only the possibility but also the need for raising such questions and for developing a framework within which they can be discussed. For there is more at stake here than the abstract possibility of the objectivity of science. We like to think of ourselves as rational, if imperfectly so; the issues concern *us*, the legitimacy of our self-images and of our conception of our relation to the world.

ACKNOWLEDGEMENTS

There are two people without whose help this book would never have been written. Alan Montefiore first suggested it and has helped it through its various stages with discussion and criticism. The greatest debt is to Jim Tiles who has not only had to live with it but has always been there to discuss, to read, to criticise and to try to sort out aberrant grammar; I owe a great deal to his understanding and encouragement. So much has been crystallised through discussion with him that the result is more ours than mine. For comments on earlier versions I am indebted to Hidé Ishiguro, Rom Harré, Eckart Förster and Mike Rosen.

ABBREVIATIONS

ARPC	*L'Activité rationaliste de la physique contemporaine*, Union Générale d'Éditions, 1977 (Paris: Presses Universitaires de France, 1951).
CA	*Essai sur la connaissance approchée* (4th edn), Paris: J. Vrin, 1973 (Paris: J. Vrin, 1928).
FES	*La Formation de l'esprit scientifique: Contribution à une psychanalyse de la connaissance objective* (11th edn), Paris: J. Vrin, 1980 (Paris: J. Vrin, 1938).
MR	*Le Matérialisme rationnel* (3rd edn), Paris: Presses Universitaires de France, 1972 (Paris: P.U.F. 1953).
NES	*Le Nouvel Esprit scientifique* (14th edn), Paris: Presses Universitaires de France, 1978 (Paris: Alcan, 1934).
PN	*La Philosophie du non: Essai d'une philosophie du nouvel esprit scientifique* (7th edn), Paris: Presses Universitaires de France, 1975 (Paris: P.U.F. 1940).
PR	*La Poétique de la rêverie*, Paris: Presses Universitaires de France, 1960.
PF	*La Psychanalyse du feu*, Paris: Collection Idées, 1949 (Paris: Gallimard, 1938).
RA	*Le Rationalisme appliqué* (5th edn), Paris: Presses Universitaires de France, 1975 (Paris: P.U.F. 1949).

ENGLISH TRANSLATIONS

The Philosophy of No: A Philosophy of the New Scientific Mind, trans. G. C. Waterston, New York: Orion Press, 1969.

The Poetics of Reverie, trans. D. Russell, New York: Orion Press, 1969, and under the title *The Poetics of Reverie: Childhood, Language and the Cosmos*, Boston: Beacon Press, 1971.

The Psychoanalysis of Fire, trans. A. C. M. Ross, Boston: Beacon Press, 1964; London: Routledge & Kegan Paul, 1964.

1

PHILOSOPHY OF SCIENCE: THE PROJECT

What are the tasks of a philosophy of science? Without an answer to this question one can hardly assess the adequacy or otherwise of any proposed philosophy of science. But how is it that there should ever be doubt about this? The professional philosopher should surely have been supplied, at least tacitly, with the answer. To engage in the philosophy of science is to enter into discussion of those problems raised in journals and books explicitly devoted to the subject. To participate in the practice is implicitly to recognise the norms of the practice, to speak the language of philosophers of science. But what now if we encounter a 'philosophy of science' which does not engage in the dialogue; a philosophy whose concern is evidently with science, but whose problems and methods resist ready identification with those which have become familiar? Is it to be judged unsuccessful because it misconceives the problems and/or adopts inappropriate methods for their solution? But on what grounds can any such normative judgement be founded? Surely our own conception of the discipline can also be subjected to scrutiny. An attempt can be made to explicate the nature of the practice in which we have been engaged. In doing so we might hope to see the relation between the two sets of concerns and thereby create the possibility of dialogue between them. We become aware that philosophical discussion cannot concern only the content of philosophies of science but must extend to consideration of the role of philosophy in relation to science and to scientists.

1 ANALYTIC ORTHODOXY

Working within the so-called analytic tradition in philosophy, one is at home with titles such as *The Logic of Scientific Discovery* (Popper 1959), *The Structure of Science* (Nagel 1961), or *The Structure of Scientific Inference* (Hesse 1974). One does not expect any unanimity in the views expressed, but one knows what sort of discussion to expect, what sort of questions will be raised, and what sort of strategies will be followed in the attempt to provide answers. The topic of general enquiry is perhaps best summed up in the English title of Pierre Duhem's book *The Aim and Structure of Physical Theory* (Duhem 1962). And indeed this work serves as a useful reference point, for besides having influenced analytic philosophy of science, it forms a part of the background to Bachelard's philosophy of science. In part Bachelard defines his position in relation to and in reaction against Duhem.

In the introduction to *The Aim and Structure of Physical Theory* Duhem sets out his project and the strategy to be followed. The aim is 'to offer a simple logical analysis of the method by which physical science makes progress' (p.3). The analysis is to proceed by first determining the *aim* of physical theory, and then, in the light of the end toward which physical theory is directed, its *structure* is to be analytically examined. Each of the parts which go to make up a physical theory is to be examined and considered in relation to its contribution to realising the theoretical goal.

This general description of the task of a philosophy of science could, with apparently minor modifications, be accepted both by Bachelard and by many analytic philosophers of science. However, this is possible only because the quite crucial phrase 'logical analysis' is given very different readings. Duhem concludes his discussion of physical theory by stressing the importance of the history of science not merely for the philosophy of science but also for the teaching of science itself. His view is that:

the only way to relate the formal judgements of theory to the factual matter which these judgements are to represent, and still avoid the surreptitious entry of false ideas, is to justify each essential hypothesis through its history.

To give the history of a physical principle is at the same time to make a logical analysis of it. . . .

(Duhem 1962 p.269, my italics)

This is one of the strands of Duhem's thought which is also to be found, much more thoroughly developed in Bachelard, even though he disagrees with Duhem's account of the way in which science progresses and hence also with his view of the history of science and of its precise relation to contemporary science and its philosophy. Bachelard is concerned with the theoretical and inferential structure of contemporary scientific thought. Like Duhem he does not sharply separate the account of theoretical structure from that of inferential structure. His concern is epistemological, with the way in which science makes progress, but it is also, more clearly than for Duhem, a concern with scientific *thought*, with the contemporary scientific mentality, not merely with its theories.

Now the idea that the history of science can be a source of logical analyses is quite alien to analytic philosophy of science, for it generally presumes that logical analysis means analysis within the framework of formal logic after the style of Frege and Russell. Thus Bas van Fraassen, in the introduction to his recent book (1980 p.2) explains his task by remarking that studies in the philosophy of science divide roughly into those concerned with the content and structure of scientific theories (foundational studies) and those concerned with the relations of a theory to the world on the one hand and to the theory user on the other (the latter will include studies in the methodology and epistemology of science). Here Duhem's concern with structure is still apparent, but there is no indication of any place for the history of science in the philosophy of science. Moreover, the study of the structure of scientific theories has tended to become separated from the study of the structure of scientific inference, from discussions of methodology and epistemology, at least to the extent that it is presumed that the first task is to form at least an approximate view of the structure and content of scientific theories by a process of logical and/or conceptual analysis. In the light of such an analysis one can then go on to a discussion of the relation of theories so conceived to the world on the one hand and to their users on the other.

On closer inspection, then, we find that Bachelard and analytic philosophers of science depart from Duhem in different directions. The apparently minor modifications to Duhem's statement of his project will be found to reflect major differences in conceptions of the task of a philosophy of science, especially in relation to the history of science. The hope here is that by bringing these differences out into the open, the focus of Bachelard's concern and the reasons for that focus will become apparent. If this is not made clear, his work is likely to be dismissed off-hand as confused because it fails to make distinctions which play a formative (and therefore also normative) role in mainstream analytic philosophy of science.

The works of Feyerabend, Kuhn, Toulmin, Lakatos and others have already made the history of science an issue. But they have also attracted the sort of dismissive criticism which does not invite further discussion of the issue, criticism which places them beyond the analytic pale for inviting an excursus into historical relativism and 'reactionary romantic excess' (Hesse 1974 pp.3–4). However the case for these works may stand, I think it would be a mistake to dismiss Bachelard's philosophy of science in this way, for even though his style is far from being formally analytic, bubbling over as it frequently does with his enthusiasm for science, his concern is very much with objectivity in science and with the rational nature of scientific progress.

a *Logic and the rational structure of science*

Duhem's influence on analytic philosophy of science is most evident in its critique of the atomistic views inherent in the original logical positivist programmes. He is credited with the holistic conception (Quine 1963 p.41) or the network model (Hesse 1974 p.24) of scientific theories, according to which scientific theories are networks of concepts related by laws in such a way that only whole theories, not individual laws, confront experience. But this influence was exerted on a way of doing philosophy of science already moulded by the logical investigations of Frege and Russell. The impact of Frege's thought on analytic philosophy has been such that Dummett could say:

Only with Frege was the proper object of philosophy finally established: namely, first, that the goal of philosophy is the analysis of the structure of *thought*; secondly, that the study of *thought* is to be sharply distinguished from the study of the psychological process of *thinking*; and, finally, that the only proper method for analysing thought consists in the analysis of *language*.

(Dummett 1978 p.458)

When thinking of the philosophy of science, we could add that from Frege we also learnt to distinguish analyses in the context of justification from analyses in the context of discovery (Frege 1959 §3), although the actual terminology is due to Reichenbach (Reichenbach 1938). That is, we learnt to distinguish between analyses of what would count as rationally adequate grounds for accepting a theorem, for taking it to be true, from analyses of the historical processes by which a person or group of people came either to accept or to propose a theory. The former yield logical reconstructions of the conceptual and deductive structures of theories, whereas the latter yield empirically descriptive accounts of historical events. Only the former are the legitimate concern of the philosopher.

The models for analyses in the context of justification (and for the subsequent rational reconstructions) were provided by the studies in the foundations of arithmetic in which Frege and Russell attempted to show that all true arithmetical propositions may be derived from definitions using logical laws alone (laws of formal logic). Even if they did not succeed in this task, they did show how numbers may be defined in terms of sets and pointed the way to subsequent demonstrations that the theorems of arithmetic can be deduced from the axioms of set theory. In this sense set theory provides a logical foundation for arithmetic. In this case it is quite clear, in the first place, that it would be a mistake to conclude that set theory either was or should be learnt before arithmetic, and in the second place, that the history of mathematics, the development of arithmetic and number theory, plays no part in the process of logical/conceptual analysis required to accomplish the task of investigating the concepts currently employed and of characterising them, as far as possible, in terms of their formally deductive interconnections with what are considered to be more basic, indefinable concepts.

On the 'orthodox' view, then, the task of foundational studies in the philosophy of science is to provide similar rational reconstructions of scientific theories, where it is emphasised that these are philosophers' artefacts, constructed with the benefit of hindsight. Such analyses are not intended as recipes to be followed by creative scientists in their efforts at theory construction. The object of the exercise is to gain a clear understanding of the structure and content of scientific theories in a way which reveals what empirical claims are made and on the basis of what primitive concepts and postulates. (See, for example, Feigl 1970.) This view has certainly not been without its critics, but even for those who do not accept all aspects of it, it still serves as a reference point for discussion.

There are several tacit assumptions here which were already referred to in the quotation from Dummett (p.5), but which deserve to be re-emphasised. (1) The philosophy of science is not concerned with the actual thought *processes* of scientists. (2) It is concerned with the analysis of the structures of scientific thought as these are manifest in the *language* of science, and more specifically as these are manifest in the *assertions* made about the physical world by scientists, whether these are proposed as theoretical claims or as reports of experimental findings. In other words, the philosopher's concern is narrowed to an analysis of theories considered as sets of statements, their logical consequences and their logical relation to reports of experimental results. (3) *Concepts* are investigated via their function in statements. (4) Logic here means *formal* logic, so that a complete logical analysis/reconstruction of a scientific theory would take the form of a formal axiomatisation of it.

b *From philosophy to philosophy of science*

There are certainly differences of opinion about exactly what formal framework can be used for such a task, about the general logical structure of a scientific theory, and about the way in which the content of such a theory is to be characterised. These are disagreements which, in part, stem from disagreements within the philosophy of language and theory of meaning. But even where there is broad agreement about the logical structure of theories, there may be sharp disagreement on issues concerning the relations of theories to the world and to their users, issues which cannot be separated from issues of metaphysics and epistemology.

The general concern of both Duhem and the logical positivists was to present scientific theories as independent of metaphysics. (Although this was for quite different reasons. Duhem was concerned to avoid conflict between scientific theorising and Catholic teaching, whereas the logical positivists were more concerned to insist on the analytic, factual and non-speculative nature not merely of science but also of philosophy itself. They shared Hume's desire to consign all metaphysical speculation to the flames.) This concern is reflected in the general division of philosophic labour in the philosophy of science remarked upon by van Fraassen, for such a division presupposes that disagreement between scientific theories will never itself be disagreement on metaphysical or epistemological issues. Philosophers of science may disagree here, and because their disagreements will be about the nature of reality and of our (possible) cognitive relation to it, this will lead to differences of opinion concerning the aim of scientific theorising and concerning the relations between scientific theories, their proposers, users and the world. These disagreements bear on the philosophy of science in general, not on the discussion of one scientific theory rather than another. Thus, for example, one finds Popper (1983) declaring the metaphysical nature of the philosophy of science. He bases his own view of science on a metaphysical realism which includes a form of Cartesian dualism of mind and matter. Equally, van Fraassen declares (1980 p.4) that he will argue for an empiricist position, which requires only that theories give an account of what is observable, and against scientific realism, which sees science as aiming to find a true description of the unobservable processes which explain the observable ones.

The debate between empiricism and realism is engaged as a philosophical debate to be settled by philosophers. A proponent of either position is required to provide (in addition to arguments for rejecting alternative views) an account of science and its language which is consistent with his metaphysical and epistemological views. This means that the task of a philosopher of science is partially determined by the philosophical, i.e. metaphysical and epistemological, starting point of the philosopher. Debate between competing scientific theories is required to be portrayed within an antecedently, philosophically set framework. The particular content of scientific theories will, therefore, never be portrayed as having any impact on the framework.

So, on this way of conceiving its task, a philosophy of science is independent of the history of science, of its actual investigative and creative practices, and of the specific content of scientific theories. Furthermore, it is concerned with theories as linguistic structures, sets of statements expressed in sentences. Theories become, in their own right, objects of philosophic investigation, and as studied are independent of their proposers and of their users. Scientific knowledge as expressed in scientific theories and as the concern of the philosophy of science thus becomes independent of the knowing subject. The epistemology of science becomes an epistemology without a knowing subject (see Popper 1972 chapter 3).

2 BACHELARDIAN HERESIES

The point of elaborating these features of the conception of the task of philosophy of science which has been dominant within the analytic tradition in philosophy is to bring out into the open the respects in which Bachelard's philosophy of science fails to fit this conception and does so because he is operating under a different conception of his task. Bachelard shares Duhem's view that the philosophy of science cannot be separated from the history of science, but rejects his claim that physical theories are free of metaphysical commitment. Like Kant, Bachelard believes that all truly scientific theories have their tacit metaphysical foundations. Moreover, Bachelard specifically denies that the way to analyse scientific thought is by analysing scientific language, so that although he largely agrees with the view of theories as conceptual networks, what he means by a concept cannot be identified with the Fregean notion of a concept.

Bachelard is concerned with scientific thought not so much in the static form of deductively interconnected sets of statements constituting scientific theories, but with the dynamic processes of correction, revision, rejection and creation of theories, with the dynamics of the experimental and theoretical practices of science. Again, his concern is not with scientific knowledge as expressed in theories, but with the knowledge, the understanding of scientists which enables them to make scientific advances. The knowing subject is never absent from Bachelard's epistemology, and, perhaps most importantly, this subject is historically located. A product, at least in part, of his culture, he is no longer the pure Cartesian intellect whose immutable rational nature is immediately transparent to him. The nature of the subject, the rational character of his thought, is mutable and develops historically. As science has developed, so have the minds of scientists. It is impossible, therefore, to attempt to give in any detail a general philosophy of science, applicable to all sciences at all periods of their history. Not only is the history of science inseparable from philosophy of science, but in addition any philosophy of science is itself historically located. In particular Bachelard sees his task, as a philosopher of science, as being to give a philosophical characterisation of contemporary, twentieth-century scientific thought and of the difference between the philosophy appropriate to classical Newtonian science and that appropriate to the science which is developing in the wake of relativity theory and quantum mechanics.

a *History and the rational structure of science*

In 1925 Whitehead gave a series of lectures in which he sought to show that the growth of science since the sixteenth century 'has practically recoloured our mentality so that modes of thought which in former times were exceptional are now broadly spread throughout the educated world' (Whitehead 1925 p.2). He went on to explain what he saw as the central characteristic of this new mentality: 'It is this union of passionate interest in the detailed facts with equal devotion to abstract generalisation which forms the novelty in our present society' (Whitehead 1925 p.3). As he said, there have in the past been plenty of practical men and plenty of dreamers lost in their wide metaphysical visions. The 'paradox' (Whitehead 1925 p.32) of modern science, or of the modern scientific mentality, is the way in which highly abstract mathematically formulated theories have been so effectively applied to practical affairs. Modern scientific thought appears to be trying to go in two opposite directions and to be doing so successfully. It insists on the importance of experiment and observation, on attention to detail and precision, making ever finer classificatory discriminations whilst at the same time looking for ever more general and more abstract theories.

This 'paradox' is the challenge which modern science presents to philosophers of science – how to understand and represent the rational structure of the interaction between these apparently opposed ways of thought. This is as true for Bachelard as for analytic philosophers of science; the overriding concern is to understand ways in which experience and theory are interconnected in scientific thought. Bachelard, like Whitehead, sees this dual concern with factual detail and theoretical abstraction as something new, as a feature of the *contemporary* scientific mentality, whereas the orthodox analytic approach tends to treat it as a permanent feature of anything which we would want to dignify with the name of science. The difference here is partly due to a different location of the philosophical problem. Neither Bachelard nor Whitehead would deny that the question of the relation between reason and experience, between theory and observation, antedates the twentieth century, but what they are struck by is the particular form in which this question arises in the context of increasingly abstract mathematical theories and increasingly precise, complex and highly instrumented experimental technology. Bachelard does not see this as reducing to the general logical problem of the relation between particular and general, or of the relation between observationally and theoretically defined terms.

Like Duhem, he sees the history of science, or rather the history of a concept or of a theory, as crucial in establishing and understanding the relation between theoretical and experimental domains. Duhem is widely known (largely via Hesse 1966) for his opposition to the idea that the models and analogies frequently employed in the development and teaching of scientific theories contribute to the content of the theory and to the meaning of its theoretical terms. Models and analogies are, for him, mere heuristic devices to be discarded once they have served their purpose. To retain them beyond this point is to invite serious misunderstanding of the physical theory through a failure to recognise the significant differences between, for example, gas molecules and billiard balls, between the domain of physical theory and the domain of common-sense, everyday experience. It is in his more detailed development of a very similar theme that some of Bachelard's most widely known terminology is coined.[1]

For, Bachelard insists, scientific thought has to break with common sense, and this transition amounts to an *epistemological rupture*. However, common sense has a tenacious hold on our thought, tending surreptitiously to infect scientific theorising, thereby setting up *epistemological obstacles*, obstacles which have to be overcome if progress is to be made. Common-sense thought deals in images, scientific thought in concepts. But where Duhem sees a smooth and continuous course of historical development for scientific theories and concepts, Bachelard sees a discontinuous, successively ruptured history. More detailed discussion of these terms and of the issues to which they relate is reserved for Chapters 2 and 4. For the present it is sufficient to note that, as Bachelard sees it, it is only via their ruptured history that present theoretical concepts receive their empirical content. At each point where a break is made with the past, when a theoretical reform occurs, the new or modified theoretical concepts have to be differentiated from, and thereby also related to, those which they are to supplant. (Again details have to be postponed until Chapter 4.) It is for this reason that we will fail in any attempt to understand the rational interconnection of theory with experience if we ignore the historical dimension of scientific theories, concentrating on justificatory, formally logical analyses to the exclusion of historical, epistemological analyses.

Does this mean that Bachelard is open to the charge of fallaciously thinking that the historical genesis of a concept determines its present content, or that the genesis of a theory is relevant to a determination of the grounds on which it should be accepted? The base from which such charges are generally levelled is one where it is presumed that any historical account will be empirically descriptive, will give merely the historical facts. The empirical contingencies of concept formation and of

[1] This has become known through the work of Althusser, who takes over Bachelard's term 'epistemological break' whilst admitting (Althusser 1977 p.257) that he gives it his own 'systematic' use.

scientific discovery, hostage to all sorts of quirks of psychology, to social and political conditions, and to sheer serendipity, can, with good ground, be claimed to be irrelevant to an account of the *rational* structure of scientific thought. However, the presumption that this is the only form which an historical account can take is one which has been questioned. Indeed, the whole difficulty of the suggestion that the history of science has a role to play in the philosophy of science is that it requires the philosopher of science to become inextricably entangled with debates in the philosophy of history. Bachelard is aware of this problem, but seeks to distance himself from it by claiming that the form of history of science that is relevant to the philosophy of science constitutes a special case, one which can be considered independently of the wider discussion of the nature and role of history in general.

He is prepared to admit that the historian of empires and nations may be required to adopt the ideal of presenting an objective record of the facts, a record in which he refrains from passing judgements and from imposing his values; he may be required to avoid subscribing to the 'myth of progress'. But for science, he says, it is different. Scientific progress is demonstrable and has been demonstrated. Moreover, the demonstration of this progress is an indispensable part of science teaching. This sense of progress is integral to the dynamics of scientific culture, i.e. current science includes within it, as an integral part, a perception of its own history, a perception which grounds its sense of the direction of its own progress. It is the job of the historian of science to describe this dynamics (ARPC pp.35–7). Thus this scientifically internal history of science is necessarily evaluative. It cannot be merely descriptive but will be required to pass judgement, to evaluate some ideas negatively as epistemological obstacles which needed to be overcome and rejected, and others positively as epistemological acts of scientific genius. What is negatively evaluated is, and is to be, excluded from contemporary scientific thought, whereas what is positively evaluated continues to play a role in science. 'This positive heritage of the past constitutes a kind of *actual past* whose action in the scientific thought of the present time is manifest' (ARPC p.36). This evaluative history is not something which can be written once and for all. On this view science is in continual dialogue with its own history. A change in current scientific thinking may well bring about a re-evaluation of past thought. By way of example Bachelard contrasts the fate of phlogiston with that of Black's work on caloric. The history of phlogiston theories is no longer the history of legitimate science (and therefore in one sense is no longer legitimate history of science) because such theories are judged to be fundamentally erroneous. The only interest which the epistemologist can have in such theories is in finding in them evidence of the kind of thought which presents epistemological obstacles. Any historian of science working on such theories has to recognise that he is working within the framework of an extinct scientific mentality.[2] Lessons for contemporary science cannot be directly drawn from such studies. By contrast, even though caloric no longer appears in theories of heat, the notion of specific heat to which it gave rise in Black's work is now a firmly entrenched scientific concept. This means that Black's work will appear as part of the approved history of science, as part of that history which is necessary to an understanding of contemporary concepts.

There is therefore a 'recurrent history' (ARPC p.38) of science, one which is continually retold in the light of the present. It is the *story* (ARPC p.39) of how scientific thinking in a specific field arrived at its present state. It is a story of the defeats of irrationality and of the progressive uncovering of the truth. The past is evaluated in the light of the present and at the same time that evaluative story contributes to present thought and to our understanding of it.[3]

The emphasis on discontinuities and ruptures in stories of the route to modern scientific theories and their concepts is

[2] Cf. FES p.11.

[3] Such a story is therefore a form of rational reconstruction, although not in the sense of Lakatos (1978) whose rational reconstructions (internal histories) are of how history *should*, conceptually, have gone judged by reference to some absolute, extra-historical standard of rationality.

what prevents them degenerating into sequences of anachronistic attributions of anticipations of modern ideas to ancient thinkers.[4] The aim of such a story is not to find our concepts already formed at some point in the past, but to reveal the route by which our concepts emerged from other concepts and ways of thought by a sequence of corrections or 'rectifications'. In this sense an evaluative history of science, or of the development of scientific thought, is not just a reconstructive tracing of the route of discovery, it also serves the function of a justificatory analysis. It is designed to show not merely *how* we came to the present views but also *why*; it reveals the reasons for rejecting previous theories, for modifying previous concepts, and thus the reasons behind the acceptance of currently accepted views.

These reasons are not psychological, sociological or political. The explanation offered is not historical in that sense. 'Reason' here means a rational (but not formally deductive) ground. The reasoning which has led to presently accepted views is therefore regarded as an important part of understanding these views, of knowing what they are. They are not only past reasons; they form part of the present justification for our theoretical positions.[5] They are active in the present via the history of science and must be understood and subjected to critical scrutiny if further progress is to be made (a view which has immediate implications for science education, with which Bachelard was very much concerned). They have to do with an appreciation of the kinds of problems that a theory was designed to solve, the extent to which it does so, and the

[4] Thus in MR, p.27, Bachelard gives a warning against venerating the past by saying that certain facts had been known to the ancients when their interpretation of those facts appears to us to have been wholly irrational.

[5] It is important, therefore, to emphasise the epistemological role of such evaluative histories. They are not in themselves to provide reconfirmations of present positions, rather they are seen as having a critical function. Unless the evaluations which are implicit in the scientific acceptance of present positions as advances over previous positions are made explicit via a recurrent history, the present position and its inherent standards will not be fully understood. It is only when we know what the acceptance of present positions rests on, what reasons can be given for counting them the culmination of a progressive sequence of corrections of past errors, that the current positions can themselves be criticised, that their justifications can be critically examined. Stressing the role of such histories in the philosophy of science does not, therefore, amount to a general advocacy of a 'whig' view of history.

reasons why, if it is correct, previous attempts were not successful and therefore had to be altered or abandoned. Such reasons do not, of course, guarantee the truth of the theory in question, but they are taken as reasons for thinking that progress has been made, for thinking that current theories represent a closer approximation to the truth than those which they have supplanted.

Here we come to the crux of the issues of the objectivity and of the rationality of science. In what sense, if any, can this claim to progress be regarded as objectively valid? Are the reasons given for thinking that progress has been made such as to ensure that it has? Are there good, *rational* grounds for accepting one scientific theory rather than another? Is there any sense in which scientific discovery or development is a rational process? These are questions with which several more recent studies in the philosophy of science have been preoccupied (for example, Newton-Smith 1981, Feyerabend 1975, Laudan 1977). Any evaluation of Bachelard's position must await a fuller account of it. But what we see, even at this stage, is that the structure of the problem is quite different depending on the context in which it is posed. The form which it takes in the context of a presumption that history has no role to play in the account of the rational structure of scientific thought is quite different from that which it takes in a context where analyses of the rational, justificatory structures of scientific thought take the form of evaluative, recurrent histories of that thought.

b *From science to philosophy of science*

A frequently recurring theme in Bachelard's works on epistemology and philosophy of science is that in order to grasp the nature of contemporary scientific thought the philosopher must be prepared to abandon traditional philosophical positions, whether these be empiricist or rationalist, idealist or realist in character. (See, for example, the Avant-Propos to PN.) The philosopher must be prepared to learn from science for 'Science in fact creates philosophy' (NES p.7). His task is not to try to fit science and scientific thought into some antecedently adopted metaphysical framework. Rather, he should derive his views on the nature of reality and on the ways in which we can have knowledge of it from an informed investigation of scientific thought.

Such a view is not, however, without its own philosophical presuppositions. It is a natural consequence of the combination of two views. (1) The Kantian view that the abstract logical and mathematical forms of rational thought play a constitutive role in experience (i.e. in relation to empirical reality), and that in this role, as forms applicable within experience and thus required to be realised within space and time, they come to determine metaphysical categories such as substance, cause and event. In other words, the formal structure of rational thought, when seen as having a constitutive role in relation to experience and to our conception of empirical reality, takes on a metaphysical aspect. (2) A rejection of the givenness and immutability of the structures of rational thought. This makes it possible to see different scientific theories and different periods in the history of science as employing different forms of reasoning, as having different and distinctive rational structures.[6] In combination with (1) this will be taken as meaning that sciences operating within different rational frameworks will also operate with different conceptions of empirical reality, and will consequently differ both in their metaphysics and in their experimental methods. This position Bachelard describes (PN p.94) as a Kantianism of second approximation.

If this is the case, then it is part of the task of a philosopher of science to tease out the metaphysical and methodological differences between sciences and between stages in the development of an individual science. It is part of his job to become aware of their various 'philosophies', and in particular to uncover the philosophies of contemporary scientific theories.

[6] These could be called 'styles of reasoning' (see Hacking 1982) or 'thought styles' (see Fleck 1979).

And he will seek to do this by investigating the forms of their theoretical thought, the structure of their experimental practices, and the ways in which these interact.

To do this he has to reveal the norms and standards operative in experimental and theoretical practices (ARPC p.17). These are what will point towards the implicit, underlying conception of reality. In aiming at knowledge of reality, our very general conception of the structure of that reality is revealed in the standards by which knowledge claims are assessed, in Bachelard's terminology, in the epistemological *values* revealed in the evaluation of such claims and in the sort of information that is taken to be important in advancing knowledge in the area concerned. Ontology is not therefore separable from epistemology. Here again the orientation of the enquiry is distinctly Kantian. Determination of the object of knowledge goes via epistemological values, i.e. via the ways in which it is presumed to be possible to have knowledge of it. This approach is summed up by Bachelard's remark (NES p.143), 'Tell me how you are sought, I will tell you what you are.'[7] The conception of an object is of something which is the source of the synthetic unity of its appearances, i.e. those phenomena which we take to be appearances of the same thing. The concept of the object is not empirically given but is a product of imposed structures, of norms, of standards of objectivity, of standards determining the epistemological value of an observation.

Bachelard distinguishes three dimensions in the standards of objectivity operative in contemporary science. These have to do with rational objectivity, technical objectivity and social objectivity (ARPC p.17). The philosopher has to try to understand the interrelation between these dimensions, to try to get at the way in which, within each of the separate sciences, these three kinds of standards work jointly to constitute the conception of the object of scientific investigation. As Cassirer put it, from the viewpoint of the critical theory of knowledge:

The epistemological exposition and evaluation of each new physical

[7] Cf. also CA p.231.

theory must always seek to indicate the ideal center and turning-point around which it causes the totality of phenomena, the real and possible observations, to revolve, – whether this point is clearly marked or whether the theory only refers to it indirectly by the intellectual tendency of all its propositions and deductions.

(Cassirer 1923 p.366)

But once it has been admitted that scientific theories may differ not only in their empirical content, but also in respect of their metaphysical and epistemological frameworks, it also has to be admitted that there can be no sharp separation between science and the philosophy of science. Science has a philosophical component. If this is so, then scientific change can require philosophic change. Thus Bachelard argues that the revolutionary changes wrought in physics by relativity theory and quantum mechanics require us to adopt a non-foundational *non-Cartesian* epistemology. They require us to organise and structure our thought in *non-Euclidean* ways, for laws are expressed using new mathematical, new rational, forms. Furthermore they require us to recognise that scientific progress does not come about inductively, that scientific method is *non-Baconian*.

The task of the next three chapters will be to explain in turn each of these aspects of Bachelard's characterisation of contemporary scientific thought. But before embarking on this, more detailed task, there is one further general feature of Bachelard's approach which requires comment, and which is indeed necessary to an understanding of the force of terms such as 'non-Cartesian', when used by Bachelard.

3 REASON AND THE RATIONALITY OF SCIENCE

As we have already stressed, Bachelard appears to violate the philosophic demand that analyses in the context of justification should be clearly separated from analyses in the context of discovery. He not merely violates this demand but requires the epistemologist to engage in a form of evaluative historical analysis which is at the same time a justificatory analysis. The epistemologist's task is not identical with that of the pure historian of science (FES p.11), but epistemology has a historical dimension. This kind of divergence in approaches to the philosophy of science reflects a deeper philosophic divergence between conceptions of reason and of rationality (reflected here in very different views of what will constitute a rational justification, and hence a justificatory analysis). Philosophers operating with different images of reason can be looking for quite different things when seeking to characterise the rational structure of scientific thought, or the rationality of science.

Analytic philosophy shares with French philosophy the tendency to treat Descartes as a father figure. Their images of reason and of the rational subject can both be seen as having Cartesian origins, but they have, in the course of philosophical evolution, diverged quite radically. Mathematics provides Descartes' model for the characteristic employment of reason in the acquisition of knowledge. But he does not equate mathematical reasoning with formal, deductive (syllogistic) inference, for he is interested in the reasoning which leads to mathematical discoveries, whereas, he thinks, formal deduction serves only as a way of organising and presenting discoveries already made. The other important feature of Cartesian reason is that it can, as in the *Meditations*, be turned in on itself in a critical fashion, asking questions about the nature and justification of its own employment. Cartesian reason is thus methodical (but not formal) in its employment and may be exercised in reflective, critical assessment.

Analytic philosophy of science has tended to focus on the idea of reason as being methodical in its employment. It has thus tended to see scientific progress as rational just to the extent that it can be seen to result from the application of methods of scientific inference and so has sought characterisations of such methods (a 'logic' of scientific discovery). Bachelard, on the other hand, tends to see critical assessment as the characteristic exercise of reason and to judge progress as rational when it is the result of such a process. The critical methodologies of Popper and Lakatos might be seen as incorporating elements of both conceptions. The question is 'What difference does the adoption of a given image of reason make to one's conception of the task of a philosophy of science?' To answer this we have to look a bit more closely at the images and in particular at the way in which they delimit the powers of reason and hence set different parameters on any discussion of the rationality of science.

a *Rational mechanism*

The pursuit of a codification of the principles of scientific inference has, in this century, taken a form which in some respects reverses the seventeenth-century rejection of the restriction of principles of reasoning to the principles of formal (Aristotelian) logic. In the seventeenth century it was reasonable to think both that mathematical reasoning was central to scientific reasoning and that this was not adequately represented within the framework of formal logic (in particular because of increasing awareness of the importance of functional relationships for science and the failure of Aristotelian logic to handle relations). But with the development of Fregean logic, designed specifically as a means of representing mathematical reasoning, it becomes at least plausible (if not beyond contention) to equate mathematical reasoning with formal, logical reasoning and to look to formal logic for a model of scientific rationality.

The image of reason which emerges is one where principles are both formal and mechanical (in an almost literal sense). Reasoning is characterised as formal in the sense that insight into and knowledge of subject-matter are irrelevant in the application of logical principles and hence in the recognition of validity. Formal principles are thus such that one can dream of an automaton, a 'reasoning machine', which would carry out inferences in accordance with them; a dream realised with the development of formal languages and computers. The modelling of any activity as rational, according to this conception of rationality, thus requires the description of a rational mechanism which would reproduce it. Thus, for example, Mary Hesse talks of constructing an artificial intelligence device, a learning machine, to represent scientific discovery (Hesse 1974 p.131).

Of course, a rational mechanism is not literally a machine, merely its software. It is, from a theoretical point of view, a mere contingency that software has to be realised in hardware, that determining the logical consequences of any piece of information takes time and is thus a process. Given the form of a statement together with the transformation rules linking it to other statements in the system, its logical implication relations are already fixed. This is a mechanism in the sense that it is a deterministic system whose structure can be described without reference to temporal processes. We are thus led away from processes of reasoning, from inference, to the study of formalised languages and the logical structure of implication relations which can be defined over them. To reason correctly, rationally, is then to follow the implication pathways already laid down in the rational structure of the language employed.

b *Reflective reason*

Descartes (as the later sections of the *Regulae* show) was not opposed to the introduction of formal techniques and computational methods into scientific reasoning. But he was insistent that reasoning, the rational thought which leads to knowledge, can never be reduced to mere computation. To go through a computation mechanically, unreflectively following prescribed rules, may lead one to correct results, but it does not yield (scientific) knowledge because it carries with it no assurance of its correctness. We can have no reason for trusting the deliverances of a computer, or any scientific instrument, such as a telescope, unless we have an understanding of the principles according to which it operates. Only on such a basis can we justify the reliance we place on it. Similarly we have no reason to trust the results we obtain merely by following prescribed rules of inference, whether they are backed by the authority of Aristotle, the Church, or are the rules sanctioned by cultural tradition, or even by our own unreflective instincts (our customs and habits). Unless we have some understanding of the basis of claims to reliability and have examined this for ourselves, we cannot claim any genuine scientific knowledge.

Here we see what is being required of a method of discovery. It is not merely to be a set of rules which, if followed, will lead to the formulation of true theories, for coming to formulate a theory which happens to be true is not to have made a discovery unless it can be recognised as such, i.e. unless it is accompanied by reasons for thinking it correct, or at least for thinking it to mark an advance over previous theories. If there is to be a method of discovery the steps of the method must themselves be the source of confidence in its results. This is possible only if, at each stage, we are aware of what we are doing and of why we are doing it, i.e. only if at the same time that the principles are followed there is awareness that they are being followed and that the situation is one in which it is appropriate to follow these principles rather than others. Only if it is possible to reflect on the course of a proof or sequence of inferential steps is it possible to assess the upshot and claim any cognitive advance as the result. This requires theoretical reasoning to be a species of practical reasoning, to be concerned with and able to reason about its own activity. It links reasoning to rational action. (See, for example, NES p.7.) The acquisition of scientific knowledge can only result from rational activity, whether mental or experimental; activity which is such that we can give reasons for what we are doing.

In the enquiry into methods of discovery we thus not only have to look at the objective conditions of the possibility of acquiring knowledge, but also at the subjective conditions. Here the concern is not with the rational structure of knowledge as the sort of thing which can be lodged in books or stored in computer memories. The concern is with reasoning with the processes of discovery, and with knowledge as a state of the individual, self-conscious rational subject.

The sort of critical reflection employed by Descartes in the first of his *Meditations*, a process of trying to doubt or call into question what had unreflectively been taken as obvious or even certain, is used as a model for the kind of rational process by which cognitive advances can be made. Bachelard labels the result of this process a 'dialectical generalisation' (PN p.137), of which more will be said in Chapter 4. It is a process of questioning and correcting both the content of our theories and the nature of our experimental and theoretical practices, our standards of justification. But since the search for secure foundations is renounced, this is not seen as a once-off questioning, as for Descartes, but as a process which has to be continually renewed. Proceeding in this way, we reject past positions only in the sense of correcting them. We move to a more general framework from which we can see the limitations of that within which we had previously been working. It is in this sense that contemporary science reveals the need to go beyond Cartesian epistemology, to a non-Cartesian epistemology. Science itself shows the need for a repeated questioning of its own foundations. Cartesian epistemology has to be corrected, generalised, but not wholly rejected, for it would be impossible to understand the new epistemology without seeing how it modifies the old. Similarly science requires us to go beyond Euclidean geometry and beyond Baconian induction. There are no rules for making such steps (NES p.12); we cannot foresee the course of scientific development. Yet critical reflection, leading to dialectical generalisation, is a paradigmatically rational process and, more importantly, it is one which appears to satisfy the demand on a process of discovery that it yield justifications as well as innovations.

c *Scientific theories and scientific thought*

We thus see that the divergent images of reason lead us to look for scientific rationality in different places: in formal characterisations of the logical structure of scientific *theories* and of the relation between these and experimental data, together with formally statable rules of theory choice and/or inductive theory development based on these representations, on the one hand; in non-formal characterisations of the epistemological structure of scientific *thought* and of the relation between experimental and theoretical *practices*, together with an account of the dynamic epistemological role of critical reflection, on the other.

As we have noted, this second approach, which treats scientific knowledge as something which only a reflectively conscious rational being can have, inevitably brings in consideration of the knowing subject and of the character of his thought processes. It would seem, therefore, that the psychologism to which Frege was so opposed is inevitable on this view and that this in itself might be grounds for rejecting any philosophical approach to science based on the conception of rationality as essentially related to the forms of justification available to the finite, spatio-temporally and culturally located subject.

There are two further objections to this approach, objections which might be raised to the idea that we should accept an account of the rationality of science based on the idea that scientific progress arises out of critical reflection. The first is that whilst such an account may exhibit the sources of a subjective sense of progress, it cannot in any way underwrite the idea that progress has actually been made, that our theories are in fact 'closer to the truth'. To reveal science as making progress in this objective sense, and to reveal this progress as the result of rational processes, we need to link these rational processes to objective rational standards, standards based on a logic whose laws are laws of truth, not laws of thought.

The second is that there is no process of critical reflection which can actually be both rational and genuinely innovative. Reflection on our own ways of thinking can possibly lead us to become aware of the rules according to which we operate, or the standards we employ. We may thus be able to describe our own, current justificatory practices and conceptual structures, but this will not afford us any basis for either justifying or criticising them. If we start from a standpoint within a culture, tradition or conceptual framework, the only standards we have available to us are those of that culture, tradition or framework. We cannot, therefore, from this basis mount a critique of those standards. If we try to mount a critique from the standpoint of another culture, tradition or conceptual framework, this will just be a matter of having opted for the standards of a framework other than our own. There can be no neutral ground where rational debate between the sets of standards can take place, for this requires rational standards which encompass all frameworks between which such debate is to take place. There is, therefore, no possibility of justifying or of criticising a self-consistent set of reasoning procedures from within, by reflection. Criticism can only come by the application of some external standard, but being external this will not amount to a rational comparison of standards, but will be a matter of the imposition of one set of standards rather than another. If there are objective, absolute rational standards these will not be ascertained by reflection on current practices.[8]

Thus, if we endorse absolute standards of rationality, critical reflection is not an objectively rational route to scientific progress. If we foreswear absolute standards, we are stuck with no rational route out of the practices in which we currently participate, and no rational route to the sort of scientific 'progress' which has required fundamental changes in our experimental and theoretical practices.

Now Bachelard is well aware that he is likely to be charged with psychologism and in several places seeks to rebut the charge. We shall have to consider these rebuttals in the course of the more detailed explication of his position. I think he also has answers to the other pair of objections. These come in the form of his grounds for rejecting the sort of metaphysical realism on which the first objection is based, and for rejecting the idea that the justificatory practices of science, or of any science at any stage in its development, will form a unitary and self-consistent set (in which case critical reflection will always

[8] This is, very roughly, the line of argument to be found in Feyerabend 1981, and is also very much the position in which Kuhn seems to leave us as a result of his analysis of scientific revolutions (Kuhn 1962).

be a potential source of modification). The structure of these responses is, however, complex and will only fully emerge in the final chapter. At this point we will also find that assessment of their adequacy as responses is far from straightforward.

2

NON-CARTESIAN EPISTEMOLOGY AND SCIENTIFIC OBJECTIVITY

When he says that contemporary science requires the adoption of a non-Cartesian epistemology, Bachelard does not merely mean that it requires the rejection of a Cartesian epistemology. He is signalling an essential connection between the new epistemology and Cartesian epistemology; for the new epistemology arises out of a critique of Cartesian epistemology in such a way that the new position cannot be understood without understanding the antecedent position and the way in which it has to be superseded. In other words, we are told that to understand his epistemology we first have to understand Descartes'.

1 CARTESIAN EPISTEMOLOGY AND THE SEARCH FOR FOUNDATIONS

Descartes, as remarked in Chapter 1, stands as a father figure to two very different philosophical traditions, traditions within which his texts are read rather differently. This is made possible by the tensions within Descartes' own position, and in particular in his characterisation of the rational subject. For there are two aspects to the Cartesian rational subject. In his search for knowledge he employs method; this makes the acquisition of scientific knowledge a discursive process; it takes time and there is reasoning involved. This is what we find emphasised in the *Regulae* and in the *Discourse on Method.* But in the *Meditations* it seems that knowledge is only achieved in clear and distinct perception (the deliverance of natural light) which takes place instantaneously, or at least involves no process of reasoning. This is intuitive (as opposed to discursive), quasi-perceptual cognition.

Are these two kinds of thought necessarily in tension? This would only be so if it were presupposed that reason, or thought, must take just one or other of these forms; that it must either be an activity or be a matter of immediate 'perception'. If this were so, the fact that we can only reach the 'cogito', the fundamental insight, by a discursive process, which is indeed an application of Cartesian method, would be highly problematic. Descartes could, however, be read not as confusedly using two different but incompatible conceptions of reason, but as in fact making the point that it is essential to our nature as finite rational beings that we have to go through discursive reasoning processes in order to acquire knowledge. But going through such processes is not enough; the mere going through the motions, the unreflective observance of rules of procedure (mechanical computation) cannot of itself result in knowledge (*Regulae*, Rule VI). The train of thought must be unified, its principle grasped. A precondition of the possibility of this is that thought not only be ordered, but that there be awareness of the order. It is not enough merely to have arrived at a conclusion (which may be correct); if we are to have any confidence in the result we must also have a conception of how we got there (*Regulae*, Rule X). If the emphasis on the reflective awareness of method is excised (as it is in the British empiricists), a quasi-perceptual reason, operating reflectively, can only reflect on and yield reflective awareness of passively received, present mental contents.

This second reading of Descartes is implicit in Bachelard's epistemological emphasis on thought processes (as opposed to the static thought contents which analytic philosophers tend to emphasise) and on the consciousness of method. It is this latter which he says is the foundation of applied rationalism (see RA p.79 and MR p.80). On this reading there is an essentially two-fold structure to the thought of the finite Cartesian rational subject, a subject capable of both intuitive and discursive thought. On such a reading, the undoubted tensions in Descartes' position will not be seen as arising out of a conflict between two incompatible models of reason (or of rational thought), but out of his attempts to find doubly secure foundations for knowledge gained by the employment of reason. Or, to put it another way, they arise out of the tension between two conceptions of objective knowledge: knowledge as understanding, based on objectively valid justifications, and knowledge of what is the case, based on a non-distorting perception. The problems arise when Descartes seeks absolute, cast-iron guarantees for the objective correctness of the insights gained by the reflective, methodical and discursive employment of reason.

Descartes' procedure is reductive (NES p.142), and the reduction is explicitly epistemological. He presumes that there are simple starting points for knowledge (epistemological foundations) and that these are found in our clear and distinct perception of simple ideas, such as the idea of extension, which grounds the self-evidence of the postulates of Euclidean geometry. Any other idea must be reduced to, or be seen in terms of, such simple ideas if knowledge concerning its object is to be possible. The explanatory ideal within Cartesian mechanistic corpuscularianism required the reduction of all physical phenomena to the motions of particles of matter and so required them to be described and understood in terms of the limited range of concepts intrinsically applicable to matter (extension) in motion.

But Descartes still needed some assurance of a correspondence between his clear and distinct perceptions and the reality of which he sought knowledge. This he could find only in the benevolence of his Creator, who could not have been so cruel as to have created him in such a way that his clear and distinct perceptions, which he is unable to doubt at the time of perception and which are therefore such that no alternative belief is subjectively possible, should nonetheless be distorted and objectively erroneous.

Cartesian science is thus grounded in an *a priori* intuition of space, the clear and distinct perception of extension which guarantees the truth of the axioms of Euclidean geometry. But with the advent of relativity theory contemporary science finds no such self-evidence in these axioms, for it has made successful application of non-Euclidean geometries. If we accept that relativity theory constitutes a cognitive advance, this move to non-Euclidean geometry requires us to recognise that what had for so long been found to be self-evidently correct, Euclidean geometry, is not thereby objectively guaranteed. The idea that scientific knowledge can find a foundation in *a priori* intuition, in clear and distinct perception, has to be called into question, for the foundations have been shown to be revisable. Moreover, within geometry itself it is no longer possible to regard geometrical notions such as 'extension', 'point', 'line' and 'plane' as simple and independently given. Even if they are treated as primitive theoretical concepts, they are, as such, defined only implicitly via their interrelations and by their relation to other concepts, relations embodied in the axioms of the theory. Concepts defined in this way are not distinct in that they are not mutually independent, nor are they guaranteed clarity; they are in no way beyond clarificatory revision. There is thus no inherent simplicity in such theoretical concepts. They are primitive from the point of view of the particular theory, but any grasp of them must be complex, for it has to be a grasp of their theoretical role, one which they play only in relation to other concepts. (For elaboration of this theme see NES pp.165–70.) This role is itself something which can be reflected upon and such reflection may itself be a source of theoretical and conceptual revision. Perfect and immediate comprehension of the content of theoretical thought is not, therefore, to be presumed. Thus Bachelard claims that innate truth does not enter into science; there are no *a priori* rational foundations. 'It is necessary to form reason in the same manner as it is necessary to form experience' (NES p.176).

Those who, like Locke, accepted both the explanatory ideals of Cartesian science and the general framework of his quest for epistemological foundations whilst remaining unconvinced by the divine underwriting of *a priori* foundations, looked not to the intellect but to the senses for their epistemological starting points. Following Francis Bacon's advice, they sought objectivity not in the abstractions of mathematics, but in immediate experience of the particular objects and events which form the starting points of scientific enquiry. But the modern scientist does not passively observe natural phenomena and does not presume that he has any direct experimental contact with the objects of his empirical investigations. In experimental science, experience of physical objects and events (observation) is both made possible by and is mediated by a complex and sophisticated technology. There are no given simples in scientific observation of this kind and no unrevisable experimental facts. Because of their indirect nature, experimental findings are always open to re-evaluation and reinterpretation.

Contemporary science thus works without the presupposition of any simple unrevisable 'givens', whether at the theoretical or at the empirical level. An account of the epistemology of contemporary science therefore has to reject foundationalist approaches, whether empiricist or rationalist. Descartes' search for absolutely secure foundations has to be abandoned, and with it the closely related idea that complete and completely secure knowledge is possible. The idea of a closed rational framework founded on simple nature has to give way to an open one, for there always remains the possibility of doubting what has been taken to be certain. This element of recurrent doubt takes us to non-Cartesianism, a sort of completed Cartesianism (see NES chapter VI).

This is a summary sketch of the way in which Bachelard motivates rejection of the foundational component in Cartesian epistemology and of traditional rationalist and empiricist approaches to the philosophy of science via an appeal to the non-Euclidean and non-Baconian character of contemporary science. But any account of the epistemology of contemporary science, and any more detailed account of its non-Euclidean and non-Baconian character, requires more than a mere rejection of foundationalism. Some positive account of what a non-Cartesian epistemology can hope to achieve is needed. Bachelard is required to say more concretely what impact the abandonment of the search for epistemological foundations has on the Cartesian project, for it is clear that he does not take this as indicating the demise of epistemology (as, for example, Rorty 1980 does).

2 CARTESIAN EPISTEMOLOGY AND EPISTEMOLOGICAL ANALYSIS

Whilst Bachelard rejects the reductive, foundationalist thrust of Cartesian epistemology, he does not reject the basically Cartesian conception of epistemological analysis. For Descartes, the strategic advantage of adopting an epistemological, as opposed to a logical, analysis is that to carry it out the analyst need only claim to be able to have knowledge of his own ideas and cognitive capacities. In the search for epistemological foundations, it only has to be presumed that by reflecting on our own cognitive capacities we can sort out which ideas are simple and which complex. The problem comes when we want to claim that what is simple relative to us and our understanding is *really* simple in the sense of corresponding to some objective, independently sustained order of things. It is by this explicitly epistemological route that Descartes hoped to be able to claim that each individual, in virtue of being a rational being, has within him the capacity to acquire knowledge for himself. Each has the capacity to reflect on and know not only the contents but also the operation of his own mind. It is on this reflective capacity for intellectual self-knowledge that the capacity for acquiring objective, scientific knowledge was (problematically) to be based. If we were pure intellects, viewing the universe from God's timeless, positionless vantage point, the order of knowledge, the epistemological order, would coincide with the objective rational order which is also the order of being. But as we are not pure intellects, epistemological ordering must take into the account the fact that we do not acquire knowledge by immediate intellectual intuition, that we have to work from what we have available to us as embodied, finite intellects in order to achieve objective knowledge (if indeed it is possible for us to achieve it). An enquiry into epistemological foundations is therefore distinct from an enquiry into either logical or ontological foundations.

It is clear that Descartes saw the imposition of an epistemological order as an integral part of his method of discovery, as part of the process of acquiring knowledge. If it is to play this role then it cannot be a purely descriptive process; it cannot be an enquiry into how we, as a matter of historical fact, have come by our beliefs and opinions. It is not, therefore, a genetic account. In this respect Cartesian epistemology is distanced from empirical psychology; Descartes does not start with the newborn child, but with himself as an adult already familiar with the world, already equipped with beliefs and opinions, but seeking knowledge. Reflective awareness, introspection, memory, may reveal how he has come by his beliefs and opinions, but only a *critical* reflection can reveal that the reasons which can be given for his having the beliefs and opinions he finds himself with are inadequate for grounding any claim to knowledge. It is only in his critical, doubting reflection that he discovers what he does not know but would like to and can ask what would suffice for knowledge in such cases. It is in this frame of mind that he can ask what sort of evidence would be needed to acquire the sort of knowledge he wants, can ask what would constitute adequate evidence.

It is this conception of epistemological analysis, of method accompanied by critical reflection, that Bachelard wishes to preserve whilst rejecting the idea that this can lead to any foundational epistemology, to any epistemology which as an independent discipline can then impose itself on science by providing the fixed framework for a philosophy of science. The twentieth-century upheavals in the foundations of mathematics and of physics force recognition that what is taken as basic in any discipline at any period cannot claim to be basic in any objective sense (NES p.179). What reflective epistemological analysis can reveal, therefore, is information about what scientific aims are and what standards of evidence (epistemological values) are operative at any given time. The results for different sciences and different historical periods may be different because the conception of the sort of knowledge which might be acquired and of how it can be acquired are conditioned both by conceptions of what is there to be known and investigated and by conceptions of human cognitive capacities.

Such conceptions are normally *implicit* in scientific practices; they are unreflectively taken on board by the participants in those practices in the course of their education. In this sense they can be metaphorically thought of as belonging to a collective unconscious. Continuing the psycho-analytic metaphor, Bachelard then talks of the epistemologist's task as a task of psycho-analysis, of uncovering and making explicit the values and metaphysical presuppositions implicit in the theoretical and experimental practices of scientists. 'We will start by asking scientists questions which are apparently psychological and little by little we will prove to them that the whole of psychology is inextricably bound up with metaphysical postulates. The mind can change its metaphysics; it cannot do without metaphysics' (PN pp.12–13). As Bachelard points out, there is frequently a gap between the scientists' perception of their practices, as this is conveyed in the 'official' explicit view, and the practices themselves. Their own, largely unreflective self-perception is not necessarily accurate (MR p.20).

In Bachelard's hands, then, the critically reflective questioning of the Cartesian method of doubt, which leads to an awareness of those things which we find indubitable and which we put beyond question, becomes an analysis of the mentality of a scientific community, revealing 'what remains subjective' even in the most rigorous scientific methods (PN p.12). It is only by making explicit the epistemological foundations of current science (its roots in the knowing subject) that they can themselves become objects of critical scrutiny, objects of critical appraisal and possible revision.

It is only if we can claim, prior to or independently of any scientific activity, to know what would constitute genuine knowledge of the natural world and to know what our cognitive capacities are that there can be an epistemology of science in general which is an independent and timeless philosophical discipline, not needing to draw on scientists' conceptions of themselves or of the reality they investigate. It is just the presumption of the possibility of such knowledge which Bachelard finds built into Descartes' philosophy in the form of an assumption that reflection on experience will bring about a clear and unproblematic separation of the knowing subject from the object of his knowledge. For Descartes, all thought is conscious thought. Reflective thought is transparent to itself in that the reflexive thinker is aware in a way which is beyond all doubt both that he is thinking and what he is thinking. There is no unconscious to be revealed.

Bachelard's relation to Descartes on this point is perhaps best brought out in his treatment (NES pp.171–4) of Descartes' discussion of the piece of wax, in the second of his *Meditations*. Here, by focussing his attention on a piece of wax, Descartes concludes:

> that perception of the wax is not sight, not touch, not imagination; nor was it ever so, though it formerly seemed to be; it is a purely mental contemplation (*inspectio*); which may be either imperfect or confused, as it originally was, or clear and distinct, as it now is, according to my degree of attention to what it consists in.
>
> (Descartes 1954 p.73)

Descartes reaches this conclusion by arguing that because the piece of wax with its colour, shape, fragrance and consistency changes in respect of these properties when it is brought near the fire and melted, these sensed qualities cannot yield knowledge of the nature of wax – that substance which persists through the change. This argument is designed to show that scientific knowledge cannot be expressed in terms of these purely sensory qualities. With this Bachelard has no quarrel. But Descartes goes on to draw the further conclusion that the mind is better known than material substance or the body. It is therefore within the subject, in his perfect self-knowledge, that all other knowledge is to be grounded. Here, Bachelard claims, Descartes assumes a radical epistemological asymmetry between the subject of experience and the phenomenon experienced, an asymmetry which is grounded in an unquestioned confidence in his knowledge of himself as a persisting substance. The continued identity of the piece of wax, as sensed, may be in question, but that of the self is not. The identity of wax as a substance can and must therefore be grounded in the identity of the self and its immediately apprehended concepts.

It is this epistemological asymmetry which Bachelard challenges, by saying that the subject changes as the wax does, as his perceptions change. He denies that this Cartesian position has any place in a philosophy appropriate to contemporary science. It is implicit neither in its methods of observation and experiment, nor in the way in which its theories are constructed. Descartes' largely passive sensory observation of the piece of wax and his wholly intellectual contemplation of this observation is contrasted with the way in which it would now be thought appropriate to investigate the properties of a piece of wax.

The wax, the object of study, would not be taken straight from the beehive, but would be as pure a sample as possible. Simplification occurs not just in the mind, but also in phenomena in the process of simplifying or purifying the object of study. Descartes, Bachelard says, fails to notice the co-ordinative role played by the experience of making something (NES pp.171–2). The wax sample would itself be the product of a systematic series of manipulations, as the product of which it is chemically defined. This wax will no longer have the smell of flowers etc. but will be an artificial product identified by its method of preparation. It must be cooled under carefully controlled conditions to ensure that the surface molecules are aligned in a manner suitable for study by X-ray diffraction techniques, for example. In other words, the very phenomena which provide the objects of scientific investigation (scientific reality) are prepared, not simply found, and the manner of preparation depends on the aspect one wants to study. Not only is the wax not known in terms of purely sensory qualities, it is not, as an object of scientific study, an object which is 'given' in pure sensory perception. The scientific study of such an object can never be complete, as each new discovery opens up fresh possibilities for investigation. The increasing complexity of phenomena is revealed in this 'progressive objectification' of both experience and thought. It involves not a subjective turning inward in an attempt to analyse what is given in perception, but a directing of thought outward in application to the physical world, attempting to create suitable objects of study, so experimentally realising theoretical concepts and objectifying them.

Many of the phenomena now studied by scientists are thus themselves the product of rational experimental methods. These methods cannot be disentangled from their theoretical embedding. 'In order to establish a determinate scientific fact, it is necessary to put a coherent technique to work' (NES p.176).[1] Theories concerning the objects of study cannot therefore be separated from technical, methodical procedures of scientific investigation. It is this which entitles the methods to be called rational. They are justified in the light of theory and consequently are also subject to critical scrutiny as a result of theoretical modifications. In this respect the observational aspect of scientific activity is essentially complex and exhibits no separation of the theoretical from the observational in the sense in which this is presumed by Descartes' disengaged intellectual contemplation on the observations yielded by his senses.

Reflection on the experimental relation of contemporary science to the objects of its empirical investigation does not, therefore, yield any clear separation of the aspects of this experience which are contributed by the theoretical (thought) activity of the scientist (subject) and those contributed by the object. Where all observation is theory-laden the functions of the intellect and the senses are neither clearly known nor clearly separable.

[1] For elaboration of this theme see Fleck 1979 chs. 3 and 4.

3 NON-CARTESIAN EPISTEMOLOGY

Reflection on modern science thus prompts a questioning and rejection of two linked assumptions of Cartesian epistemology: the assumption that we should be looking for epistemological starting points, indubitable foundations, and the assumption of the possibility of privileged and complete self-knowledge on the part of the reflective knowing subject. The possibility of complete self-knowledge is one of the assumptions needed to underwrite the idea that in empirical knowledge the subject's contribution can be clearly and unproblematically separated from the contribution made by the object.

If it can be presumed that the medium through which one is looking is perfectly transparent and non-distorting, there is no problem in sorting out what part of what is seen is due to the medium and what to the object viewed through it; it can be assumed, with the naive realist, that things are just as they are seen to be. If it cannot be presumed that the medium through which one is looking is perfectly transparent and non-distorting, one cannot extract from what one sees information concerning the shape, size, colour or surface features of the objects viewed through it without learning about the characteristics of the medium. Only when these are completely known can one know what part of what is seen is to be attributed to the medium and what to the object viewed. Once this information is available it may be possible to reconstruct the object viewed theoretically. But this requires that there be some way of finding out what the characteristics of the medium are. The Cartesian assumption of the complete transparency of thought to itself similarly allows us to reconstruct a wholly objective view of the external world. It is in this way that the subject can, at the outset, drop out of the epistemological picture once attention is on objective knowledge of the external world.

But if everything had to be viewed through a distorting medium, if there were no independent route to knowledge of the medium's characteristics, then the problem of sorting out what it contributes to what is seen could only be solved indirectly and gradually. It would require a mutual adjustment of theories concerning the objects viewed and concerning the medium through which they are viewed. (This problem is not merely illustrative, but genuinely arises in a slightly more complicated form in connection with the use of instruments such as microscopes, whether optical or electron. In both cases the slides have to be prepared. As a result there is room for controversy concerning how much of what is seen is a product of the preparation techniques used. Such controversy can only be resolved by reaching an agreed theoretical account of what effects the methods of preparation have, and so of how they contribute to what is seen. This in turn requires a theoretical account of the material being prepared, but then the microscope was being used precisely to find out more about this material.)

Thus, if the Cartesian assumption of the possibility of privileged and complete self-knowledge on the part of the reflective knowing subject is called into question and suspended, the separation of subjective from objective in empirical knowledge becomes problematic. The subject whose knowledge acquisition is in question can no longer drop out of the epistemological picture and cannot, even for theoretical purposes, be treated abstractly as a pure intellect with an objective, if partial, view of the world. The separation of intellectual, purely cognitive concerns from concerns arising out of consciously registered interests and from values having origins not fully accessible to consciousness is part of what will have to go on in the acquisition of objective knowledge. Knowledge has to *become* objective. Not only are there no starting points, no epistemological foundations for natural science, there are no already fully objective views, right or wrong, about how the world is.

It is from this perspective that one sees why Bachelard rejects any purely philosophical theories of epistemology and metaphysics. Traditionally they are conceived as theories which, if correct, provide the framework for all discussion of the development of science. Any such framework either explicitly provides, or builds in implicitly, a conception of the nature of reality (the object or objects of objective knowledge) and a conception of the knowing subject and his cognitive capacities. But such a theory, having claims to be timelessly applicable, is possible only if it is allowed that any philosophically minded person can attain a view of himself in his capacity as a knowing subject and also of the kind of things of which he could expect to have knowledge, and that he could attain this in such a way that he could have some assurance that his view was objectively correct and applicable to all knowing subjects at all times. Without Cartesian self-knowledge this is not a possibility, for there will be no such time-independent account to be had, even from a God's eye view, let alone from the vantage point of a historically located person.

But Bachelard sees the implications as extending further. The necessarily co-ordinated development of conceptions of subject, object and their relation is not simply a historical development of opinions about some fixed items standing in a determinate but not fully known relation. It is at the same time a development both of the knowing subject and of the object of his knowledge (NES pp.177–80). The subject's conception of himself-in-relation-to-his-world partially determines his identity as a knowing subject, for it shapes the way in which he acts, thinks and seeks further to define both himself and that world. The detailed features of his rational cognitive capacities cannot be disengaged from his conception of those capacities, and this conception is shaped by and revealed in the rational discourse of his culture. It is only the reflexive character of the rational subject, his potential for reflective, critical self-judgement, which is presumed to be present throughout, if not always exercised. Equally, by his planned intervention in the world, intervention which is governed by his conception of the world with which he is interacting, the world investigated by science changes. New phenomena and new environments are created.

Non-Cartesian epistemology, as a 'dialectical generalisation' from Cartesian epistemology, must thus suspend the assumption that there are epistemological foundations and the assumption that subjective and objective are unproblematically separable in empirical knowledge. It must provide a more general, more abstract epistemological framework within which, by the addition of different assumptions, we can regain not only Cartesian epistemology (the epistemology of Cartesian science) but also the epistemology of Newtonian science, or quantum mechanics, or micro-biology.

4 NON-CARTESIAN EPISTEMOLOGY AND THE REJECTION OF REALISM

At first sight the suggestion that an epistemology of natural science, the pursuit of objective knowledge of the external world, should do without the assumption that the knowing subject is clearly separable from the object of his knowledge seems problematic. It has seemed to many that this leads straight into relativism, instrumentalism and abandonment of the view of science as a rational discipline concerned with the acquisition of objective knowledge. This, for example, is the view expressed by Popper (1983), whose defence of realism presupposes that the existence of an independent material world possessing in itself a determinate (but not deterministic) structure, independent of us or of our thought about it, is a necessary condition of the possibility of rationality and objectivity in science.

As Popper recognises, quantum mechanics presents a *prima facie* challenge to this view. He seeks to meet this challenge by arguing that quantum mechanical theory can and should be interpreted as descriptive of a world of particles, particles which have fully determinate positions and velocities even if it is in principle impossible for simultaneously accurate measurements to be made of both. It is just this form of realism to which Bachelard objects. It is the realism of classical physics, in which objective truth is ultimately grounded in the states and intrinsic qualities of independently existing objects (particles).

It is a position which Bachelard occasionally describes as '*chosiste*' (NES p.42, for example).

The problem with the term 'realism' is that it can and has been used to denote many very different positions. Bachelard is not opposed to all of those positions which have been labelled 'realism', but he is specifically opposed to this realism with respect to things. As he sees it, quantum mechanics and relativity theory require the abandonment of this view and so force a reappraisal of what is meant by objectivity in science. It is no longer possible to presume that the objectivity of scientific knowledge is grounded in the natures of independently existing *things*. The substance–attribute metaphysics needs finally to be discarded. It is now a question of a 'realism of second position, a realism in reaction against the usual reality, ... of a realism made of reason realised, of reason experienced' (NES p.9).

The challenge which has to be met by this rejection of objectual realism as a basis for the account of objectivity in science is that of giving an alternative account of objective knowledge, one which does not ground this in correspondence between scientific descriptions of individual objects (particles) and the intrinsically and independently possessed states of those objects (particles). Like Popper, Bachelard sees science as aiming at objective knowledge. But unlike Popper he is prepared to admit that at different periods of history, and within different scientific communities, scientists may be operating with different conceptions of this goal, i.e. have different views as to what it would be to have objective knowledge of the particular phenomena they are investigating.

A change in scientific practice may involve methodological changes whose legitimation requires a changed epistemology and metaphysics, a change in views of scientific knowledge and of its means of acquisition. If such a change is judged to be progressive, it has to be allowed that a change in conceptions of scientific knowledge can also be seen as constituting progress. In other words, progress has to be allowed to be possible not only in science, but also in its philosophy. Bachelard requires philosophy to progress beyond the thing-based realism of classical physics.

But then we have to ask, what are the benchmarks of progress? How can there be scientific progress if there is no single, shared conception of the goal of science, no shared set of standards relative to which progress can be assessed? When we are further denied the philosophic right to assume the existence of absolute standards and an absolutely neutral vantage point for in-principle evaluation, what right have we even to the *concept* of scientific progress? Bachelard's answer, briefly, is that objective knowledge, as the limit and goal of all scientific enquiry, together with its co-ordinate conception 'reality', is a purely functional or formal notion. It has a constant role as the source of regulative principles, of standards of evaluation, even though its positive content is subject to revision. Progress in science is progress *towards* objectivity. Such a brief answer obviously stands in need of elucidatory expansion.

5 THE STRUCTURE OF AN EPISTEMOLOGICAL FIELD

To understand the way in which a concept, such as objective knowledge, can play a constant role in relation to the goal of scientific enquiry across the most radical theoretical and philosophical revisions, it will perhaps be helpful to take a slightly less abstract example, that of a chemical element (cf. MR pp.73–9). When chemical elements are defined as simple substances, i.e. substances which cannot be further resolved or analysed by chemical means and which are the ingredients of all other substances, they are defined in a purely functional or schematic way. The definition not only does not fix the list of chemical elements, it gives no criteria for making this determination. To do this we would first need to know what things count as substances and what are the possible chemical means of decomposing substances. It is this lack of specificity which makes it possible for the idea of an element to play the same role in Aristotelian and alchemical element theories, in Lavoissier's fundamental reorganisation of chemical concepts, and in Mendeleev's construction of the periodic table of elements.[2]

What will be counted as an element is relative to agreed methods of chemical analysis. The delimitation of this class of operations is in turn relative to chemical knowledge, technology and the conception of substances in general and their identity criteria in particular. But what is common to all approaches in which the notion of an element is given employment is a particular view of the general form to be taken by explanations of chemical phenomena. It will be a framework in which an account of the nature of a chemical substance and subsequent explanations of its observed properties will draw both on its structure and on its material composition. It is an explanatory structure which is both analytic and countenances irreducible differences between (simple) substances. (This framework, it should be noted, has not had universal acceptance. Boyle, in common with mechanistic corpuscularians, was amongst those who rejected it. For them all differences between substances were reducible to differences of structure imposed on the 'universal and Catholick matter'. See Boyle 1979 chapter II for example.)

In other words, the use of even such a schematically defined notion already points to, because for its filling out it must be located in, a field of theoretical enquiry structured in a particular way. That the concept is employed at all allows us to say something about the theories in which it is employed, something which is independent of the specific content of those theories. Moreover, the fact that elements are defined in relation to methods of chemical analysis gives rise to possible sources of criticism for any particular theory of chemical elements, for its account of elements and of chemical operations and methods of analysis must form a coherent whole. If

[2] For example Boyle says, when arguing against element theories 'I now mean by elements, ... certain primitive and simple, or perfectly unmingled bodies; which ... are the ingredients, of which all those called perfectly mixt bodies are immediately compounded, and into which they are ultimately resolved' (Boyle 1911 p.187). An account based on the periodic table goes: 'Elements are those substances which go through chemical manipulations without being resolved into simpler structures. They make up the distinct varieties of matter of which the universe is composed' (Davis 1952 p.1).

what is said to be an element appears to be analysable by a method recognised as a chemical operation, or if substances known to be composite resist analysis by recognised methods, some aspect of the system requires revision; it cannot be regarded as wholly adequate. It was on these grounds that Boyle rejected both Aristotelian and Paracelsan element theories, arguing that the products of fire analysis were, in some cases, further decomposable, and that some compounds could not be reduced to their supposed elemental constituents by fire analysis. The discovery that water is decomposable into gaseous components, and that it can be reformed by their combination rules out any theory in which water is regarded as an element. It played an important part in the recognition of gases as substances and hence in the chemical revolution initiated by Lavoissier. In addition to this, the recognition of gases contributed to a refutation of the idea that fire is a universal analyser by making it possible to see the change which occurs in the heating of metals as one of combination (the formation of a metallic oxide) rather than analysis. With the change in the status of this operation there also goes, therefore, a change in the status of metals; they could no longer be regarded unquestionably as mixed substances.

The account of the objects of chemical theory (substances and their elemental constituents) has to be co-ordinated with an account of the methods used for their manipulation and investigation and so with the epistemological significance which is attached to certain operations (those classed as analyses) as revealing the constitution (nature) of a substance. But why, or under what conditions, should one think that changes in chemistry resulting from this kind of criticism have been progressive? There are several dimensions along which one might try to ground this judgement; however I shall take just one which Bachelard stresses (although not to the exclusion of all others).

Bachelard's suggestion is that to get a sense of the historical progress of chemistry one should focus on the development of conceptions of the purity of a substance (theoretical accounts of substances, standards of purity and purification techniques).[3] Such conceptions are deeply embedded in the theoretical and experimental structures of chemistry and yet the claim to have been able to obtain a purer sample of, say, gold by some new means is one which would generally be thought to be justifiable, even though this is a claim to have made progress of a certain kind. In a similar way claims to increased accuracy in measurement are regarded as justifiable, even though (1) the techniques required for that increased accuracy will be highly theory-dependent, and (2) judgements of increased accuracy cannot be based on comparison with any absolute standard. Such judgements have, ultimately, to be based on the coherence of theoretical and experimental practices in which the measured quantity is involved. This is the case also for judgements as to the purity of a substance. 'The *purity* of a substance is thus the work of man. It should not be taken for something given in nature. It retains the essential relativity of human works' (MR pp.79–80).

In many ways Bachelard's account of science represents a generalisation from this kind of example. The example serves to illustrate how a schematically characterised goal can structure a cognitive field and thereby introduce an order which overarches particular theories within it without requiring any absolute yardstick for the imposition of that order. The order arises out of internal comparison between successive theories. The ideas of closer approximation to purity and to accuracy provide instances of the more general idea of approximation to objective knowledge. Better standards of chemical purity and of accuracy in measurement go hand in hand with progress toward objective knowledge. This kind of progress cannot occur merely as the result of adopting new theories, but nor can it occur wholly as the result of experimental innovations. In this respect it illustrates the point that standards of scientific objectivity involve both theory and experiment; both rational and empirical standards are involved. The structure imposed on the epistemological field constituting the physical and biological sciences by their goal – objective knowledge – is that of a

[3] See NES p.19 and MR pp.71–81.

continuing dialogue, or a dialectical play, between theory and experience. It is with the nature of this dialogue, with its dialectical structure, that Bachelard is, at various levels of detail, concerned in his works on the epistemology of science (NES p.18, for example).

6 OBJECTIVE KNOWLEDGE

To become clear about the order being imposed when Bachelard sees science as concerned with the acquisition of objective knowledge, we should note what it rules out. Bachelard is equally opposed to instrumentalist accounts of science, on the one hand, and to the mystico-romantic strands inherited from *Naturphilosophie*, on the other. His rejection of these views of contemporary science is inseparably linked with the revised account of scientific objectivity required by a non-Cartesian epistemology as one in which the quest for epistemological foundations is renounced and in which the separation of subjective from objective is not taken as given. Non-Cartesian epistemology requires an abstract and schematic characterisation of objective knowledge, one which can be specialised in a variety of different ways. Bachelard does not explicitly provide such an account, but does implicitly employ one.

The touchstone of his conception of objective knowledge lies in its opposition to what is subjective. There are two dimensions to this opposition (NES p.15). (1) Minimally one can say that objective knowledge of a given domain (the object or objects of knowledge) is such that the correctness or otherwise of knowledge claims is determined by the nature of that domain itself. There are two slightly different ways in which this idea can be further explicated: (a) it must be possible to make sense of the idea that a person claiming such knowledge could be mistaken, (b) *complete* objective knowledge is impossible; however much is known there always remains something unknown.[4] These are both attempts to capture the idea that

[4] 'What is belief in reality.... It is essentially the conviction that there is more to an entity than is immediately given' (NES p.34).

there has to be room for a cognitive distance between the knowing subject and the object of his objective knowledge. In this sense objective knowledge is essentially a two-term relation. (2) Since the correctness or otherwise of claims to objective knowledge is determined by reference to the domain about which the claim is made, the knowing subject should ideally abstract from himself as far as possible in making such claims. Judgements as to what is objectively the case should be the same for any rational subject placed in perfect cognitive conditions and where all individual biases introduced by non-cognitive concerns are set aside. This is not to say that every item of objective knowledge should, in principle, be accessible to every individual rational subject, regardless of his historical, cultural or psychological context. These contextual factors may be thought to make it impossible for a particular individual ever, even in principle, to be placed in ideal conditions for making that particular judgement. The idea is much more like that embodied in the principle of universalisability as applied to moral deliberation. It is designed to sort out the kind of reasons which can be used in the justification or criticism of claims to objective knowledge; they must be reasons for belief which do not depend merely on characteristics of the individual concerned. This is summarised by appeal to the idea that they should have a universal cognitive value; i.e. that they would weigh with any other rational being in the same cognitive situation. Here there comes to be an essential link between rationality and objectivity; claims to objective knowledge are required to be rationally justifiable. Exactly what account can be given of the rational justification for belief and of what can be counted as 'the same cognitive situation' are problems to which individual epistemologies address themselves.

Objective knowledge may thus be said to be non-subjective in two senses: (1) it is such that the object of knowledge (what is known, whether this be a particular thing, phenomenon or fact) is distinct from the knowing subject at least to the extent that there is room for there to be a cognitive gap between them; (2) it is independent of the individual, non-cognitive constitution of the subject, so that it is the possible knowledge of a depersonalised, rational subject. Thus any individual aiming to acquire such knowledge must both become aware of and seek to eliminate possible sources of error, the ways in which or the reasons why he might make mistakes in his cognitive judgements. At least in his scientific life, he must aspire to conform as closely as possible to the theoretical ideal of a purely rational subject. It is by combination of (1) and (2) therefore that we get the idea that objective knowledge is the product of critical rationality.

The seventeenth-century attempts to provide epistemological foundations for the new mechanistic science, either in the clear perception made possible by the natural light of reason or in immediate sense perception, can be seen as operating on the assumption that the objectivity and the objective correctness of knowledge are guaranteed when the subject's cognitive state is wholly determined by the 'object' perceived and is therefore to be found in those cognitive states in which the knowing subject is most passive, i.e. in immediate sense perception or in immediate intellectual intuition. The two different locations of epistemological foundations correspond to the two senses in which knowledge may be thought to be non-subjective and hence objective. Whereas empiricists sought objectivity in immediate contact with the individual objects of empirical knowledge, rationalists sought it in contact with the divine mind, in a pure rational intuition (closely akin to mystical or religious experience) of the most abstract and general ideas on which the objective rational order was thought to depend.

This same association of objectivity with subjective passivity can be found in more recent philosophers who continue to employ a basically perceptual model of objective knowledge. Thus Mackie, when motivating his form of moral scepticism, argues: 'If there were something in the fabric of the world that validated certain kinds of concern, then it would be possible to acquire these merely by finding something out, *by letting one's thinking be controlled by how things were*' (Mackie 1977 p.22, my italics). Mackie clearly thinks that the existence of objective values is necessary to grounding the objectivity of moral judgements and that these would have to be known ('perceived') by a special faculty of moral intuition (Mackie 1977 p.38). His argument provides an illustration of the way in which the realist's ontological grounding of objectivity in independently existing items is bound up with the epistemological grounding of objective knowledge in the presumed passivity of the subject in immediate perception, on a quasi-perceptual intuition.

We should therefore expect to find Bachelard rejecting (as indeed he does) this epistemological grounding of objective knowledge in presumed perceptual passivity. Indeed we have already seen (pp.32 and 37) the way in which he emphasises the non-passive character of experimental science. Moreover, Descartes had already, with his Evil Genius, raised the question of the possible subjectivity and unreliability of the deliverances not only of the senses, but also of the natural light of reason (his clear and distinct perceptions). Might not the Evil Genius have so created him that his intellectual vision was irremediably distorted? Irremediably because when he contemplated the ideas directly he could not doubt, they determined his thought. But what passive, unreflective thought found irresistibly compelling, reflective thought could question. It is thus in the activity of reflective thought that reasons, justifications and criticisms are given. Even for Descartes, the objectivity of his knowledge is not assured at the level of immediate intuition, but only at the level of discursively rational, critically reflective thought.

It is this aspect of Cartesian epistemology which Bachelard's non-Cartesian epistemology preserves, whilst turning away from all reliance on intuition, on purely experiental knowledge. Immediate, intuitive thought becomes, for Bachelard, the source of epistemological obstacles, the source of assumptions, of ways of thinking about and looking at the world which have to be discarded in the quest for objective knowledge. What is found to be intuitively obvious, what can unreflectively be taken for granted, is seen as reflecting the nature of the subject. In such unreflective thought he is *qua* critical, rational subject indeed passive, but to that extent his view is conditioned by his non-rational nature, a nature formed in part by the culture to which he belongs and in part by the universal values of the human soul or psyche, values which operate below the level of conscious thought.

Instrumentalist accounts of science insist that science has, and can only have, the prediction and control of phenomena as its goal. Scientific progress is thus demonstrated by an increased capacity to control the environment and to predict occurrences in it successfully. Science is valued for its utility. The grounds for holding that this is the only possible goal for science, for denying that it can aim at objective theoretical knowledge, are those derived from empiricist epistemologies which presume objectivity at the level of what is immediately given in experience. It is presumed that there is a level of phenomena relative to which predictions can be objectively tested. The very characterisation of the instrumentalist position thus presumes a subject–object separation, a separation of theory from objectively testable predictions, of the subject's ways of thought from the given realm of phenomena, a realm presumed to be objectively knowable. Bachelard's criticism of such presumptions, developed in his critique of empiricism and of objectual realism, can thus also be used to rule out instrumentalist accounts of contemporary science (see RA p.6). Such accounts ignore the active role of reason and of rational organisation in the experimental aspects of contemporary science.

Romantic idealist philosophies of nature, which had a certain currency in the nineteenth century, lie at the other extreme, ignoring instrumented experiment altogether and recognising no demand for systematic empirical testing of theories (RA p.5). The appeal is to a different, otherworldly form of immediate experience. Bachelard complains that such philosophies never go beyond an 'ethereal sensualism' (RA p.6) working only with images, grounding objectivity neither in intersubjectivity nor in empirical observation, but in a form of intuition which involves a kind of intellectual sympathy by which one places oneself within an object in order to coincide with what is unique in it. Such knowledge is therefore strictly inexpressible; it can only be approximated in metaphor, by the use of images. The attainment of this kind of knowledge involves a complete identification of the subject with the object; it is knowledge which has to be experienced. It is on this that its claim to objectivity (non-subjectivity) is based. Bachelard insists that on the contrary scientific objectivity does not require a loss of identity of the subject in union with an object, but a loss of individuality. It is an objectivity with an essentially social dimension; objective knowledge is not the unique experience of an individual but that which could be agreed upon by all similarly placed rational subjects. It is seen as an objectivity achievable only by a self-consciously reflective subject, conscious both of its cognitive capacities and of their limitations. It is only to the extent that the individual subject is conscious of his own contribution to his perceptions of things that these can be put aside, as they must be in objective judgement. Scientifically objective knowledge is therefore, for Bachelard, always reflective, discursive knowledge and never immediate, intuitive (purely experiental) knowledge (cf. NES pp.14–15).

7 SUBJECT–OBJECT

This demarcation between intuitive and discursive knowledge is as crucial to Bachelard's view of science as the demarcation between science and pseudo-science is to Popper's. The two demarcations do not, however, coincide, for Bachelard not only rejects the kind of realism which Popper advocates, he also rejects the idea that science is continuous with common sense (whereas for Popper the growth of scientific knowledge is the growth of ordinary human knowledge *writ large* (Popper 1963 p.216)). The pivotal distinction in Bachelard's contrast is not between the empirically testable (falsifiable) and the empirically irrefutable, but between subjective and objective. Around this distinction there cluster a number of contrasts:

image–concept, soul–mind, intuitive–discursive, common sense–science. That there should be such a cluster of contrasts here reflects Bachelard's conviction that there cannot be a unitary approach to man and his knowledge (MR p.19), for he has an essentially dual nature (an intellect and a soul). These dual aspects are equally important and although distinct are seen almost as mirror images of each other. Thus the importance of the qualitative uniqueness of an experience is not denied by Bachelard; its significance is, however, subjective rather than objective.

Bachelard's account of the subjective realm owes much to Jung's depth psychology.[5] This is particularly clear in the way in which he dismisses alchemy and all other theories based on the four elements Earth, Air, Fire and Water, as not even representing the first stages toward a science of matter, but as presenting an obstacle to any such science (MR p.57). Element theories of this kind will not even be part of the pre-history of chemistry; they will not be mentioned in the sanctioned histories of science which trace the development of our current chemical concepts. This is because on Jung's interpretation alchemical thought has a wholly subjective origin. According to Jung alchemy is not a theory of matter, or even seriously an attempt to turn base metal into gold; it is a projection of the unconscious. It is a mystical, religious way, a sequence of symbols and symbolic acts which have a ritual, psychological function. Its doctrines do not, therefore, reveal anything about the nature of the material world. They are, however, a source of insight into the nature of the 'archetypal contents of the collective unconscious' (Jung 1968 p.32). In Jung's opinion, the soul has its own religious values although it (and they) may be stuck in an unconscious state. The soul is not comprehensible to the conscious mind; it cannot be the subject of discursive knowledge. Of alchemy he says: 'alchemy is rather like an undercurrent to the Christianity that ruled on the surface. It is to this surface as the dream is to consciousness, and just as the dream compensates the conflicts of the conscious mind, so

[5] See, for example, MR pp.26 and 48–54.

alchemy endeavours to fill in the gaps left open by the Christian tension of opposites' (Jung 1968 p.23). It is with dream worlds and dream images, as revealing the structures of the unconscious, that Bachelard associates alchemy (see PR pp.75–7). It is for this reason the antithesis of science and belongs with poetry. It is a voyage of the subject into itself, into the 'universal basis of the individually varied psyche' (Jung 1968 p.33) and is thus a surpassing of the individual empirical subject but not in the direction of the world of material objects. It is a voyage that needs to be taken, but not under the illusion that its end will be objective (scientific) knowledge. It needs to be acknowledged for what it is, a source of insight into the self; the union sought is not unity with an external object, but unity within the subject, a form of self-knowledge (cf. MR p.55). Indeed, it was with the role of Fire and fire images in subjective, imaginative thought that Bachelard himself (in PF) embarked on his own philosophic passage into the realm of poetic, imaginative and essentially subjective thought.

It is not only alchemy, many of whose practitioners, although not all (it is dangerous to generalise over such a long and varied tradition), explicitly aimed at a non-rational, mystical form of knowledge, which Bachelard finds guilty of failing to be science. All theories based either on the four elements, or on the three principles (Salt, Sulphur, Mercury) are similarly condemned. Theories organised around the four elements were not confined to the alchemical, mystical tradition; they also, in conjunction with the four humours, formed the Aristotelian framework of the common-sense medieval view of man and the world, shaping medical theories and practices. (Bachelard gives an example from Agrippa on pp.44–7 of MR.) This common-sense life of the four elements derived largely from Aristotelian and Galenic rational, discursive conceptions of knowledge. If such element theories are unscientific it is not because they explicitly espoused any mystical conception of knowledge. What is true of such element theories is that they are not empirically testable. It is part of the Aristotelian view that here on earth there can never be a pure sample of

Earth, Air, Fire or Water. Everything is composed of all four elements, the proportions in its elemental composition being the cause of the variation in the observed characteristics of different substances. In consequence, a distinction has to be drawn between the water found in rivers, where the element Water predominates, and the true element Water. The latter is not observable. And, Bachelard says, 'To be verified by *everything* is a way of escaping all verification' (MR p.51).

This serves to illustrate two points concerning the relation between the views of the science–non-science distinction as presented by Bachelard and Popper. The first is that it becomes clear that Bachelard does require a scientific theory of matter to make contact with experience of the material world in such a way that theoretical revision may be required in the light of experience. A theory, such as the Aristotelian theory of elements, which does not do this is wholly *a priori* in origin, and this, for Bachelard, means that its origin is in the subject. The elements are projections, or illegitimate objectifications, of subjective values (MR pp.59–60). The second is that it is this kind of example which serves to support his claim that science is (necessarily) discontinuous with common sense. The 'sciences' of matter had to turn their backs on medieval common sense in order to become sciences. The determinants of common-sense thought are such that there is no basis for presuming this to be, or to be aiming to be, objective thought about the world. Bachelard does not presume that man is naturally, or indeed ever, a purely rational animal. The Aristotelian framework which had become the framework of common-sense thought presented an epistemological obstacle to the development of a science of matter.

Thus the fundamental epistemological break which has to be made for science (or the scientific mind) to come into being is a break with what is given as intuitively self-evident, or obvious, whether this be at the level of ordinary sense perception, everyday experience, or abstract principles. What is obvious has to be regarded as open to question. To value what is self-evident because it is self-evident is to reflect and to project subjective values. What is found to be self-evident is a function of the psychological make-up of the knowing subject, who cannot be presumed to be dispassionately rational. To make this break is thus to begin to break with the subjectively conditioned viewpoint. And it has to be a break, a reversal of values, for the values of the conscious rational mind are opposed to those of the unconscious (MR p.50). The unconscious is not educable whereas a mind which values rationality is eminently educable. Objectivity demands that what has always seemed to be self-evidently true be questioned. The self-evident, highly valued in conceptions of knowledge in which the paradigm is perceptual, is devalued by science.

The term 'epistemological rupture' is also used, by Bachelard, in connection with the development of science itself. This development is seen as discontinuous, being punctuated by epistemological breaks. But a break in the history of science does not involve the same wholesale reversal of epistemological values that is required to move from common sense to science, or from alchemy to chemistry. It is, rather, the result of a change in the epistemological value attached to some particular belief or cluster of beliefs, which, having been taken for granted, are called into question. Because such changes occur within a science, i.e. within a framework in which objectivity is valued, such changes are seen as an essential part of the rational process which constitutes the development of science.

Thus, although the objective knowledge sought by science is rational, discursive knowledge, there is nonetheless in the historical course of scientific development a continual interplay between intuitive, experiental (subjective) and rationally discursive (objective) forms of knowledge. Pure objectivity is neither given nor ever fully achieved, but has to be worked for and worked towards. The dialogue between experience and theory is thus accompanied by a dialogue between subjective and objective modes of thought. The importance of the continuing interplay between intuitive and reflectively, rationally discursive thought for the dynamical structure of scientific thought will become clearer in Chapter 3, but already begins to emerge with a consideration of the place of the subject in non-Cartesian epistemology.

8 OBJECTIVITY AND THE NON-CARTESIAN SUBJECT

We have just seen that for Bachelard an epistemology of science does not start with the newborn child and a consideration of how he comes by his concepts and his beliefs. It starts with the transition to science, with the emergence of science from common sense, from the unreflective subject already fully equipped with concepts, beliefs and language. Because the development of scientific thought requires a break with common-sense thought (whatever the historical period to which the scientist belongs), scientifically objective knowledge can never be 'first knowledge', it is always reflective knowledge, is the product of reflection on previously held beliefs and ways of thinking.

Further, we have also seen how it is impossible for the subject to drop out of a non-Cartesian epistemology of science once the assumption of the possibility of complete self-knowledge is dropped. Increased objectivity and increased self-knowledge now have to go hand in hand. But beyond this, the emergence of science, of scientific thought, requires that the scientist consciously adopt objective knowledge as his goal, that he recognise and seek to satisfy the demand for and standards of objectivity. The conception of what he is striving for has to be internalised.

For example, it is not sufficient that the experimental scientist make an observation report which may (or may not) be factually correct, and which he merely thinks to be objectively true or false. He needs to be able to rely on his observations and needs to be able to justify that reliance both to himself and to others. Positivist attempts to ground experimental science in the incorrigibility of reports on what is immediately observed were designed to justify such reliance. But the recent history of many sciences shows that experimental results frequently have to be re-evaluated, and even when not re-evaluated there is a recognised demand that they should be repeatable and checkable. Present standards of objectivity are such that there must be the possibility of correction even at the observational level of science.

Thus an epistemological account of the experimental base of science cannot rest wholly on the scientist's capacities to make perceptual discriminations. The whole point about such recognitional capacities is that they are identified as capacities (reliably) to acquire immediate (non-inferential, intuitive) knowledge of a given kind. If one makes mistakes, one lacks the capacity for (correct) recognition. In response to this situation it has seemed to some (for example Popper and Newton-Smith) that a strong dose of realism is what is required; assume the independent existence of the object being observed and do not build success into the exercise of perceptual recognition capacities. In this way it can be allowed that there is the possibility of mistake even in the exercise of such intuitive capacities. But this would have the scientist picture himself as playing a game of chance with nature without any prospect of increasing his chances of winning (making more accurate observations) by developing his observational skills, for when a cognitive faculty is presumed to be immediate and intuitive there can be no hope of providing further criteria for sorting out the cases in which mistakes are made. The situation in which mistakes are nevertheless admitted to be possible is very much the situation in which the Evil Genius places Descartes in respect of his clear and distinct perceptions. The result is a scepticism which only a benevolent God can overcome.

This provides further support for the suggestion that it is wrong to try to discuss the experimental aspect of scientific epistemology in perceptual, observational terms. Even at the level of measurement, there is no thought in the mind of the scientist that his instruments yield data which are in any way absolute and beyond further correction or refinement. New instruments will no doubt be invented and new techniques developed. When such thoughts are made integral to the 'observational', experimental activities of scientists, it becomes clear that they are not engaged in simply registering what is (objectively) the case. It is crucial that methods of data collection be recorded as well as the results themselves. No scientific journal would accept an account of experimental results alone. The experimental techniques used must be included so that others can scrutinise and evaluate the methods in addition to attempting to repeat the experiments with a view to checking the results.

The so-called N-ray affair (see Bloor 1976) provides an example. In 1903 the French physicist Blondlot published papers in which he claimed to have discovered a new kind of ray (N-rays). Several other French physicists confirmed his results and continued to investigate the properties of these rays. But the American physicist R. W. Wood failed to repeat Blondlot's results and on visiting Blondlot's laboratory to see the experiments for himself subjected the experimental techniques to severe critical scrutiny. He himself did not observe the phenomena which the French physicists claimed to be present. In response Blondlot published (1905) an elaborate set of instructions on how to observe N-rays. It was essential to avoid all straining of vision, to avoid any conscious fixing on the luminous source whose variation in brightness one was trying to observe, and so on. It was admitted that this required practice and that some people might never acquire the ability.

To insist that experimental observation requires the development of special observational skills, which not everyone may be able to acquire, is, in itself, correct. Problems occur only when, as in this case, all attempts at indirect, instrumental confirmation fail, so that the only evidence is perceptual and hence highly dependent on the 'sensitivity' of the individual observer. In this situation the phenomenon is left irremediably subjective.

A new experimental phenomenon, if not theoretically understood, is already known under a publicly accessible description: the intentional description derived from the design, construction and execution of the experimental set-up in which it was observed. To engage in experimental activities, the scientist has to know what he is doing; he acts according to a rationally designed plan: 'modern science is founded on the *project*. In scientific thought, meditation on the object by the subject always takes the form of a project' (NES p.15). Before stating his results publicly in a scientific journal the scientist has to reflect on his experiments, his reasoning in carrying them out, and on their actual execution. In other words, he must be able reflectively to evaluate his activity from a standpoint which is, as far as possible, that of the impartial rational subject. It is achievable to the extent that, as a member of a scientific community, the scientist is aware of its standards and can thus foresee the kind of critical appraisal to which his experimental claims will be subjected.

Thus the scientist's conception of what makes for reliable observation will be based on a recognition that his observation reports will be subject to the critical scrutiny of other scientists and in his knowledge of the kind of critical standards they will apply. The subjective grounding of scientific objectivity at the level of experiment does not, therefore, require a retreat into the privacy of sense perception which is immune to criticism because inaccessible to others. Rather, it requires the reverse, namely a minimising of reliance on the discriminatory powers of the senses in favour of grounds which in relying least on the limitations and peculiarities of the individual observer and in using the most sophisticated techniques available are likely to take into account most of the possible sources of criticism available to other scientists. Thus being able to view oneself, one's reasoning and actions as if from the standpoint of others is a necessary condition of that form of critically reflective self-consciousness which marks the transition to scientific thought, thought which has internalised the demand for objectivity.

Yet this seems, on the face of it, to be no more than a social, cultural grounding of objectivity, a reduction of objectivity to intersubjectivity, in which case the experimental, empirical base of science will remain culture-relative. Different groups of scientists may operate with different standards, with different conceptions of the demands of objectivity, in which case it seems that (in the absence of the strong realist commitment) there is no basis for any rational debate across such differences. This is roughly the situation in which the arguments of those like Kuhn and Feyerabend leave us, and which has led them to say that since the idea that science has an absolutely objective observational base is no longer tenable, we must give up the view of science as capable of making cognitive progress, as capable of increasingly objective knowledge. For standards of objectivity are, they argue, culture-relative. In Feyerabend's work the result is an onslaught on the whole conception of science as a rational enterprise concerned with the acquisition of a form of knowledge distinct from mystical and religious insight. For he claims that the scientist is in the grip of an ideology; he is deluded if he thinks that he can, by his methods, or any other, acquire objective knowledge. The whole enterprise rests on a chimera. The scientist is really no better off than the alchemist or the astrologer, and he is more self-deceived because he believes himself to be superior to them. The romantic, poetic, aesthetic values emerge as the only ones to have genuine application.

Kuhn, by contrast, whilst denying that one can talk of scientific progress as a rational progress toward objective knowledge, falls back on an instrumentalist notion of progress: progress in manipulation and control, increased predicative power and ease of use of theories (Kuhn 1970 p.261). However, Kuhn would agree that scientists rarely see their enterprise in wholly instrumentalist terms. The argument is rather that this is the only sense in which there can be talk of progress; a revolutionary change, one which alters socially accepted standards of objectivity, cannot be rationally grounded at the cognitive level, but can be grounded at the practical, instrumental level.

Again we have already seen (p.52) that Bachelard will not accept the possibility of such a retreat into instrumentalism. From his point of view the mistake made by both Feyerabend and Kuhn is that they fail to consider the subjective conditions of objective knowledge. They treat the scientist, *qua* rational subject, as wholly formed by his cultural setting, as unable to transcend it other than by a pure act of choice, he is portrayed as unable to deploy any critical standards other than those which his scientific education has instilled in him (cf. Chapter 1 pp.25–6). Bachelard's argument is that when we realise that the demand for objectivity has to be internalised by individual scientists, that this requires them to be able to view themselves and their actions from the standpoint of others, we also realise that it is this same ability which is a precondition of mental privacy and which thus confers a degree of mental autonomy (RA ch.IV). It is the condition of the possibility of deceit, of the concealment of thoughts, feelings, emotions and desires, and thus of self-consciousness as consciousness of an individual distinct from and not wholly constituted by society. It is a knowledge of how mental states are observed, awareness of the methods and criteria used by others, which creates the possibility of concealment. Thus it is only against a background of general publicity that this sort of mental privacy is possible. Concealment can have no point if there is no possibility of revelation or discovery. To the extent that this form of privacy is the mark of mental autonomy, it is a result of the general autonomy of an agent capable of intentional action and of evaluating that action, an agent conscious both of himself and of others. But as with any observational knowledge, the agent's self-knowledge may be far from perfect, and will be capable of improvement.

Mental autonomy and its conscious recognition by the self-conscious, reflective subject is important, for in this autonomy lies the germ of transcendence of culture and hence of the possibility of grounding a conception of objectivity which is not culture-bound. With the internalisation of this autonomy comes the possibility of internal recognition that the standards of objectivity employed in science are imposed standards, that objectivity cannot be taken as given even within the cultural context. It is for this reason that the development of the autonomous subject in the poetic expression of his private dream-images, an expression which breaks the bounds of language, is as important as the intellectual development of the scientific mind. The conditions for creative science are not wholly separable from those of creative artistic work.

In proposing a new theory or a new set of experimental techniques it is necessary for an individual, or group of individuals, to think outside the accepted norms, to break with them. But for scientific development this cannot be just an arbitrary break; it has to be rationally, critically motivated. It has to arise out of a questioning of the correctness or legitimacy of accepted experimental or theoretical practices. The proposers of any radically new approach must expect to have to make out their case to fellow scientists. This can be an appeal to reason only if reason is not itself wholly culturally constituted and if other scientists have themselves internalised this possibility, the possibility that in principle questions might be raised about accepted standards. This requires scientists themselves to regard the framework within which they are working as open, as having a potential for development, rather than as fixed, final and closed and therefore the framework within which all future research must be envisaged as proceeding. It is just such an attitude on the part of scientists which Bachelard claims to characterise contemporary science. It has internalised the possibility of change and thus works within a rational framework which is, in the above sense, open (cf. NES p.179).

It is here, then, in the complex interplay of subject–object, subjective–objective, that Bachelard's response to the relativists' criticism (Chapter 1 pp.25–7) of the rationally innovative powers of critical reflection lies. It is the structure of this interplay which thus has to be examined in more detail. What we see is that for Bachelard the escape from relativism rests on the delicate balance of the relation between subjective insight and discursively rational thought, between the autonomy of the individual and his own recognition of the essentially social nature of his being as a scientist, as a member of a scientific community. It is on the instability in the balance of this relation that two potentially relativising restrictions are played off against each other as scientific thought seeks to break the bounds of both. By reference to the demands of socially applied standards of objectivity, relativisation to the individual standpoint is overcome, but it is the autonomy of the individual, his potential for breaking with socially accepted standards that overcomes relativisation to the standpoint of a particular culture (cf. NES p.15). And in both cases relativisation to the pure subjectivity of idealist thought is to be overcome by recognition of the demand for an experimental base – theory has to engage with experience, not in the passivity of observation but in the activity of the experimental project.

The problem is to see how this sort of case can be made out in more detail. It is very similar to the problem which arises in connection with Wittgenstein's views on the nature of mathematics (Wittgenstein 1967). These views stand accused of radical conventionalism because Wittgenstein insists that the discovery and acceptance of a putative proof as a proof constitutes a mathematical development in the sense that it marks a change in the concepts involved and hence also a change in mathematical thinking. This means that he does not see the grounds for the acceptance or rejection of a mathematical proof as already given in the accepted meanings of the terms used to express the proposition which is to be proved. The difficulty is to say how one can both take this view and insist that the acceptance or rejection of a proof is not a matter of arbitrary, free choice, but is something for which reasons can be given, i.e. is something which can be rationally grounded. It is no accident that we should find an analogue of the problem at the level of mathematics, for it is in mathematics that Bachelard finds his model for rational thought in general and for scientific rationality in particular. It plays a crucial role in his account of creativity within contemporary science; mathematics provides the realm within which the scientist can daydream. It thus plays, for Bachelard, a crucial role in the subject–object mediation.

3

NON-EUCLIDEAN MATHEMATICS AND THE RATIONALITY OF SCIENCE

One of the most striking features of the two theories which revolutionised physics at the beginning of this century (relativity theory and quantum mechanics) is the extent to which they are mathematical theories. For this reason, no discussion of the philosophy of contemporary science can ignore the role of mathematics in science. Any such account will have to say something about the way in which mathematics is applied in physics and as such will be based on, or will tacitly presuppose some view of the nature of, the (pure) mathematics being applied.

Bachelard sees the role of mathematics in contemporary scientific thought as extending beyond the organisation and expression of particular theories to the provision of frameworks for rational thought which reach outside those theories. This is to accord mathematics a central epistemological role in science, a role in theory construction and development. Mathematics, Bachelard says, provides the space within which scientists dream: '. . . the poetic art of Physics is done with numbers, with groups, with spins' (PN p.39). He further holds that these mathematical, 'anagogical' reveries must be sharply distinguished from ordinary reveries, those which form the subject-matter of depth psychology. Grasping this distinction is, he says, crucial to understanding the psychology of the scientific mind. But just why is this so crucial?

It is because, as we have just seen (pp.62–5), at the growth point of scientific theorising the active role of the subject has to be reconciled with the demands of objectivity. Conceptual innovation requires creativity on the part of the subject, who asserts his autonomy in thinking beyond accepted norms. But if this conceptual innovation is to be even a candidate for cognitive advance, the subject has not merely to be able to communicate his critically constructive thought, he has also to be able to convince others that there are reasons for taking it seriously. His scientific reveries, although his in so far as they are not dictated by, but in some respects go against, what is generally accepted, have nevertheless to be candidates for objectification, and so have to be objectively assessable. Objective and subjective have to meet; it is the subject's autonomy that makes possible any objectivity beyond intersubjectivity and yet this free creative thought has to be capable of achieving intersubjective acceptance; it has to be the thought of a rational subject, not of an individual; it has to be rational thought.

Is the emphasis on mathematics as the medium of scientific reveries likely to show how this is possible? One reason for thinking that it might is that it is within the philosophy of mathematics that the problem of objectivity is posed as one of determining whether innovative theoretical work is a matter of creation or of discovery, and if a matter of creation of how nonetheless there can be objectivity in mathematics. It suggests that if, or to the extent that, sense can be made of the idea that there is cognitive progress in mathematics, this might be used as a model for the progress of science along its theoretical axis, progress in its rational component.

It is over the nature of mathematics and of mathematical thought that Bachelard differs most radically from analytic philosophers of science. His view of mathematics is very much more Kantian in spirit than theirs. Although he spends very little time in explicit discussion of issues in the philosophy of mathematics, a position on such issues is implicit throughout his works on the epistemology of science, given that mathematics plays such a key role in shaping his view both of the structure of scientific thought and of its dynamics. Thus when he describes contemporary scientific thought as non-Euclidean, this does not simply mean that modern scientific theories make use of non-Euclidean geometries; it means that the ways of thinking which are characteristic of more recent mathematics, and which are symbolised by the development of non-Euclidean geometries, have been incorporated into scientific thought. Thus to understand the difference between classical and modern physics as well as the nature of the transition from one to the other, it is first necessary to understand the ways in which mathematics has changed and continues to change. Then, in order to see how this affects science, one needs an account of the role of mathematics in science.

1 NON-EUCLIDEAN GEOMETRY AND THE DEMISE OF GEOMETRICAL INTUITION

Prior to the successful application of non-Euclidean geometry in relativity theory, it was still possible to regard Euclidean geometry as a pure, abstract mathematical theory which is descriptive of physical space. Even after the mathematical discovery of the possibility of non-Euclidean geometries, it was still possible to regard Euclidean geometry as having a special status, for these others were mere conceptual possibilities without application. They did not therefore contain 'truths' or embody knowledge in the way in which Euclidean geometry did. Thus even Frege still regarded the axioms of Euclidean geometry as expressing *a priori* truths and was totally opposed to Hilbert's treatment of them as implicit definitions of the terms 'point', '(Euclidean) line', '(Euclidean) plane', etc. (see Frege 1971 pp.6–21). Euclidean geometry, in its employment in classical physics, thus appeared as a body of knowledge grounded in *a priori* truths. For realists about space the Euclidean axioms were true in the sense of correctly describing structural characteristics of space, and our ability to have *a priori* knowledge of their truth had to be predicated on our possession of an intuitive faculty (geometrical intuition) by means of which we are able to recognise such features of physical space. For Kantians the privileged position of Euclidean geometry was grounded in the fixed forms of our intuition. The axioms of Euclidean geometry have application in the world of experience because they correctly describe the structure we impose on that world, they correctly characterise the space which is the form of our outer sense. Our ability to have *a priori* knowledge of the truth of these axioms is then grounded in our capacity for reflective self-knowledge, and in particular in our capacity to construct geometrical images (such as that of a dimensionless point tracing out a line of no thickness) in the mind in abstraction from all physical conditions (in our capacity for *a priori* construction in pure intuition).

With classical physics these two kinds of view, both grounding the privileged position of Euclidean geometry in geometrical intuition, but of different kinds, were still a possibility (although other, more nominalistic positions were of course also possible). But with the successful application of non-Euclidean geometry in physical theory, these were no longer possible views (cf. Bachelard's criticism of Poincaré: NES pp.40–2). We have already seen how Bachelard uses this in his wider argument to the effect that within contemporary science it is no longer possible to suppose that objective scientific knowledge can be grounded in intuitive, perceptual or quasi-perceptual, recognitional capacities. But what is now at issue is mathematical rather than scientific knowledge. Here too Bachelard sees contemporary standards of objectivity as having moved away from reliance on any special faculty of mathematical intuition.

His understanding of the move away from intuition, although in part motivated by the above arguments, is in fact much more closely based on mathematical developments, and in particular on the arithmetisation of analysis. For geometrical intuition was under mathematical attack long before the development of relativity theory. From the very introduction of calculus, its infinitistic methods and its use of infinitesimals were regarded with suspicion. It is all too easy (as Berkeley was quick to point out) to be led into paradoxes and contradictions when calculating with the infinitely small. Geometrical intuition, based on visual images, fails us when we try to extend intuitively grounded geometrical reasoning beyond the perceptible to the infinitely small (CA pp.170–2).

The motive force behind the arithmetisation of analysis was the desire to eliminate reliance on geometrical intuition, for geometrical intuition was manifestly not a secure basis for developing analytic geometry incorporating the infinitistic methods of calculus. And yet it was necessary, in the absence of any reductive account of continuous spatial extension (the continuum), for the continuity of spatial extension to be taken as a primitive given. A continuous whole has to be treated as one which is given before its parts, because it cannot be constructed out of points. This is true even on Kant's account, for the synthesis which yields an *a priori* intuition of something continuously extended, such as a line, is a synthesis in intuition, it is not a rational synthesis but one which involves imagination in the construction of a mental image or representation (*Critique of Pure Reason* A102). From a rational, conceptual standpoint, continuity is primitive and the direction of mathematical investigation takes the form of an analysis, a decomposition of a continuous whole.

With the work (in the nineteenth and early twentieth centuries) of Wierstrass, Dedekind, Cantor and Hilbert, this has changed. The continuum has a structure; it can be treated as being composed of (non-denumerably many) points. The continuity characteristic of Euclidean space is no longer a simple primitive concept, but a highly complex one. A continuous space is no longer constructed in the imagination but is rationally constructed by laying down axioms. The treatment of space and of functions of continuous variables no longer has to be analytic. The direction can be reversed to that of a rational synthesis. It is possible to define and make use of functions which are everywhere continuous but nowhere differentiable. The graph of such a function cannot be drawn; in this sense it is a purely rational construct. The mathematical means of characterising structure have ceased to be representational in any pictorial, or picturable, sense.[1] The transition from geometry to algebra initiated by the use of Cartesian co-ordinates and the methods of analytic geometry has been completed with the move to point-set topology and the theory of real numbers.

But the mathematically successful completion of this foundational project raised rather than resolved philosophical questions. What is a rational construct and by what means can we have knowledge of it if it is not picturable and is not accessible to geometrical intuition? Is there some other faculty of rational, mathematical intuition required? If mathematics deals only in constructs is it right to think of mathematical knowledge at all? These were the issues which received much discussion during the first half of this century. Bachelard only explicitly addresses these, by now traditional, issues of the philosophy of mathematics in his earliest work, *Connaissance Approché*. At other times he is careful to distinguish between questions of objectivity and epistemology as they occur in mathematics and in empirical science and confines himself to the latter, whilst at the same time emphasising the extent to which mathematics structures the rational thought of scientists.

Despite this acknowledgement of the distinction between mathematics and empirical science, there are common themes in Bachelard's treatment of the two domains. Most notably these are (1) rejection of intuition as a basis for objective knowledge, and (2) insistence on the link between conceptions of the object of knowledge and the route of epistemological access to it. The term which he applies to his own view of mathematics is 'constructivism'. Constructivism is, to be sure, a

[1] Hardy, for example, says: 'The notion of a derivative or differential coefficient was suggested to us by geometrical considerations. But there is nothing geometrical in the notion itself. The derivative $\varphi'(x)$ of a function $\varphi(x)$ may be defined, without any reference to any kind of geometrical representation of $\varphi(x)$, by the equation

$$\varphi'(x) = \lim_{h \to 0} \frac{\varphi(x+h) - \varphi(x)}{h}$$

and $\varphi(x)$ has or has not a derivative, for any particular value of x, according as this limit does or does not exist' (Hardy 1952 p.213).

broad church, but its various sects do share the view that, ultimately, mathematical objects are our creations and that the development of mathematics in a new direction is not a voyage of discovery into pre-existing but uncharted lands. It is, therefore, a rejection of mathematical realism. But equally, since constructivists do still want to countenance talk of mathematical objects as constructs, they reject both the logicist reductions of mathematics to logic and the formalist treatment of mathematics as simply a body of calculating techniques, devices for facilitating inference in other subject areas. Constructivists do, however, differ among themselves over the nature of the creative, constructive process, that which brings mathematical entities into being. In this respect Bachelard's position differs in an interesting way from that of many constructivists.

2 FORMAL LOGIC AND THE AVOIDANCE OF PSYCHOLOGISM

The Scholastic view of geometry (based on acquaintance with Euclid's *Elements* and Aristotle's *Posterior Analytics*) was of a deductive science, of a body of knowledge in which the axioms and postulates are self-evidently true and where all subsequent propositions are logically deduced (in principle by the use of formal, syllogistic inference) from the axioms and postulates together with definitions. Here, in theory, the only point at which there need be appeal to intuitive knowledge is in relation to the axioms and postulates. But with the geometrical innovations of Descartes, Newton and others, it became clear that geometrical reasoning was quite unlike anything which could be syllogistically formalised. The idea that mathematics has its own, not strictly formal, modes of reasoning based on an intuitive grasp of the subject-matter concerned therefore became reasonable and had wide currency. Geometrical intuition was thus appealed to as revealing not only the truth of Euclid's axioms but also for grounding the non-formal, non-logical, but nonetheless *a priori* and deductive character of geometrical reasoning.

However, the developments in formal logic made by Peano and Frege, and consolidated by Russell and Whitehead, coupled with the arithmetisation of analysis, made a return to a more Scholastic view of mathematics and of mathematical reasoning possible. For the new logic was capable of formalising arithmetical and algebraic reasoning, but not traditional geometrical reasoning. Russell reflected this shift when he said:

> The whole doctrine of *a priori* intuitions, by which Kant explained the possibility of pure mathematics, is wholly inapplicable to mathematics in its present form. The Aristotelian doctrines of the schoolmen come nearer in spirit to the doctrines which modern mathematics inspire; but the schoolmen were hampered by the fact that their formal logic was very defective. . . .
>
> (Russell 1917 p.74)

The search, on the part of Frege, Russell and others, for a more adequate formal logic was intimately connected with their conception of what was required to put mathematics on an objectively secure footing. Their concern for the foundations of mathematics was not primarily epistemological, not with how *we* come to know the truth of mathematical propositions, but with how their truth can be proved; not with how *we* reason mathematically, but with how we can be sure that the use of mathematical reasoning and the application of mathematical results will not lead us astray, that it will not allow us to deduce false conclusions from true premisses.

The very framing of this concern presupposes a rejection of the Kantian position; it is a concern which can only arise on presupposition of a clear separation between scientific statements, which are objectively true or false, and the mathematics which is to be applied in the deduction of further scientific statements. Empirical content is presumed to be separable from mathematical form. From either the realist or positivist points of view from which this kind of question about mathematics can be raised, the objectivity of mathematics, the objective security of its scientific use, will be secured only by showing that the application of mathematics does not in any way distort or contribute to the empirical scientific subject matter to which it is applied. This will be the case if mathematics can be treated as a language which it is convenient for scientists to use but whose use is in principle eliminable either via a series of definitions (as would be the case if Russell's logicist reduction treating all mathematical objects as logical fictions, or logical constructs, had succeeded) or via a direct axiomatic formulation of physical theories (see, for example, Field 1980). In other words, mathematical reasoning should, if possible, be reduced to logical reasoning and this in turn should (1) be grounded in objective relations between statements on the basis of their truth conditions (the way in which they relate to the world) not on the basis of the way in which individuals understand them or on the way in which either individuals or people in general make inferences, (2) be formal, i.e. be carried out according to laws which hold in virtue of form not content, since reasoning should transmit content undistorted. This would mean that no intuitive knowledge of the nature of the subject matter would be required for valid inference, merely a knowledge of the logical forms of the statements involved and of the formal laws of correct inference. The total elimination of the psychological, of all reference to the subject and his cognitive capacities, in the system of linguistic representations which constitute scientific knowledge was seen to be the only route to securing objectivity, and thus of making objective knowledge possible. (Cf. Frege 1979 pp.2–8.) The whole direction of this thought rests on acceptance of the idea that subjective passivity is a necessary condition of objectivity. All thought of reasoning as rational activity, as possibly constructive, and as a possible route to discovery (which is a transition in the *subject's* cognitive state) is thus required to disappear. Thus we have Russell saying:

In the discussion of inference, it is common to permit the intrusion of a psychological element, and to consider our acquisition of new knowledge by its means. But it is plain that where we validly infer one proposition from another, we do so in virtue of a relation which holds between the two propositions whether we perceive it or not: the mind, in fact, is as purely receptive in inference as common sense supposes it to be in perception of sensible objects. The relation in virtue of which it is possible for us validly to infer is what I call material implication.

(Russell 1937 p.33)

(Russell goes on to explain that p materially implies q if and only if p is false or q is true.)

Such a reduction of reason to a wholly passive role, whether of mechanically following formal rules or of the receptivity of an intellectual intuition, is totally rejected by Bachelard. He admits that given the ideal of axiomatising mathematical and scientific theories there are two ways of seeing scientific knowledge as wholly untainted by psychologism: by adopting realism or formalism (RA p.27). The rejection of realism (outlined above on pp.42–4) will extend to both mathematics and logic. For even one who adopts a realist stance in relation to logic and meaning still has to face epistemological questions. How do we get to know about what statements (objectively) mean and what their logical relations to each other really are? Russell's appeal to a form of intellectual intuition merely displaces the role of intuition from geometry or arithmetic to logic; it does not eliminate it.

An alternative to realism is to take a formalist, or a conventionalist attitude. Logical forms are forms of our language, forms instituted by convention and intersubjectively grounded in linguistic practices. No further justification or grounding is either possible or required. This alternative is equally firmly rejected by Bachelard. He criticises formalist approaches to axiomatic theories in general, whether the theories in question are scientific or mathematical. In his view they make the mistake of trying to collapse method into habit by mechanising reason, when method is in fact the antithesis of habit.

Axiomatic systems are mathematically constructed psychological robots (RA p.25). But reasoning is absent from the functioning of robots. Formalism, by attempting to mechanise reason, is in danger of itself sliding back into a position from which it is impossible to claim any objectivity in respect of its formal rules. For it is in danger of degenerating into an automated rational organisation, one employed wholly unreflectively out of habit. In logic, psychologism is identified with the view that the laws of logic are laws of thought, sanctioned by the nature of thought, not grounded in its objects. Frege's rejection of psychologism was in part an insistence that logical laws, as instruments in the justificatory adjudication of claims to objective knowledge, must be laws of truth, not laws of thought; they must be laws grounded in the objects of knowledge, propositions or the contents of assertoric sentences. But if logical laws are reduced to the level of habit, or accepted practice, to linguistic convention, they have (and can have) no justification, no objective grounding (see, for example, Dummett 1973a). They are not subject to critically reflective appraisal, but become laws definitive of a wholly conventionally constituted rationality. Thus, if opposition to psychologism was designed to rescue the possibility of objective knowledge, conventionalism threatens a return to psychologism in a new guise (collective psychologism) for it offers only intersubjectivity as a substitute for objectivity; laws of language take the place of laws of thought.

We thus return to the issues raised at the end of Chapter 2. Bachelard argues that it is necessary to put a little psychology back into the formulae of formal logic in order for a genuine non-psychologism to develop in conscious opposition to the psychological. It is not enough that one reason in accordance with logical laws, there needs to be an exercise of reason; it is necessary to become conscious of the demands of rationality *as* rational demands, as part of the demand for objectivity. The subject's consciousness of his own cognitive activities cannot be dropped out of an account of scientifically objective knowledge because it forms an essential precondition of the possibility of objectivity. Objective knowledge is not just a matter of true belief, of factually correct assertion (see pp.58–61). It requires a grasp of the reasons for belief or assertion. This is why justification cannot be separated from discovery. Bachelard expresses this by saying that scientific thought has a double aspect, assertoric and apodeictic (normative). (See RA chapter 2.) Similarly there has to be consciousness not only of the fact of assertoric thought, but also of normative thought, consciousness of rational organisation, of standards and methods of reasoning. Even thought within a formal system, thought according to rules, requires more than mechanical rule-following. Calculation is in the first place used to reach a result, but to be sure of the correctness of the result one must be able to go back and check the calculation. The calculation and its result have to be evaluated; if correct the calculation is seen to *prove* its result.

Pure formalism is to be rejected even in the case of geometry. The possibility of non-Euclidean geometry may rule out any Platonistically realist attitude toward Euclidean geometry, but this does not mean that geometrical thought must therefore be seen as reducing to the empty formal framework suggested by Hilbert's axiomatisation. The rational framework of geometry has to be epistemologically more 'engaged' than that. Any system of axioms has to be applied; it has to be used. It takes intelligence to see how to make use of an axiomatised theory, how to use the axiomatic formulation as an instrument of clarification. In the case of geometry, this requires, amongst other things, awareness of the relations between different geometries. Logical formulation in mathematics is necessary for rigour, but it reflects only one aspect of mathematical thought. It omits the less tidy processes of mathematical discovery. (A case persuasively argued in Lakatos 1976.) These are the processes which have to precede axiomatisation.

Axiomatisation is only possible by reflection on a preceding activity. It is a *re*ordering of past thought (a *re*thinking, never a first thinking) in the interests of rigour. This is why, for Bachelard, rationalism is something which has no beginning. Axiomatisation, or the imposition of rational order, is a normative, evaluative reordering. It can only occur in the presence of a consciousness of lack of rigour, in rectification of which a rigorous, systematic order is imposed. Axiomatic thought thus has a double aspect (its formal character and its character as a formalisation of non-formal thought) which is masked by its presentation as a simple formalism. For this reason, study of the logical foundations of a kind of knowledge can never exhaust epistemological study of that knowledge. So logicism and formalism are both to be distinguished from rationalism of the kind which Bachelard advocates.

3 ARITHMETIC – REASON IN ACTION

It is worth dwelling on the divergence between Bachelard and the Frege–Russell–Hilbert-inspired formal approach to the philosophy of mathematics, for it has a profound effect on subsequent options for an account of the role of mathematics in science, and in mathematical physics in particular.

If it is assumed that objectivity requires passivity on the part of the subject in cognition, so that intuitive, recognitional capacities are the only possible basis for objective knowledge, and if it is further accepted that all mathematical reasoning is reducible to formal, logical reasoning, then it seems impossible both to deny that intuition is the source of objectivity in mathematics and to admit the possibility of any kind of objective mathematical knowledge. Thus Russell said:

> In short, all knowledge must be recognition, on pain of being mere delusion; Arithmetic must be discovered in just the same way in which Columbus discovered the West Indies, and we no more create numbers than he created the Indians. The number 2 is not purely mental, but is an entity which may be thought *of*. Whatever can be thought of has being, and its being is a precondition, not a result, of it being thought of.
>
> (Russell 1937 p.451)

For the possibility of objective knowledge then appears to rest upon the independent existence of mathematical objects. But these, as abstract objects, could only be known by a mathematical intuition, a faculty playing a role analogous to that of sense perception in relation to physical objects. It is this traditional Platonist attitude toward mathematics which is placed in question by the recognition that reliance on intuition leads to contradictions and paradoxes. The mathematical rejection of intuitive clarity in favour of formal rigour (which Bachelard endorses) seems then to lead ineluctably to a nominalistic and ultimately formalistic denial of the possibility of mathematical knowledge.[2]

But, as we have seen, Bachelard, even whilst stressing that eliminating reliance on intuition is part of the conception of objectivity embodied in contemporary mathematics, equally rejects both logicist and formalist positions. (Indeed these views do not accord with mathematicians' own perceptions of their activity and have been most strongly opposed by those who have worked within mathematics.) This makes Bachelard's position puzzling; he appears to reject all the options. But this is because he rejects the basic assumptions behind all foundationalist approaches to the philosophy of mathematics. He rejects both the assumption that mathematical reasoning is reducible to formal logical deduction and the linking of objectivity with subjective passivity. His remarks about the rejection of intuition concern the *standards* (ideals) operative within mathematics; they are not to be confused with statements concerning the actual status of mathematical theories. As remarks about standards they do not entail that there is no place for intuition in the account of the nature of mathematical knowledge. Quite the reverse, for as representing one of the concerns which motivate research and one of the standards applied in assessing progress, they presuppose that there is pre-existing mathematics to be rigourised.

In associating mathematical thought with activity on the part of the rational subject Bachelard's approach is Kantian in spirit, even though it departs very significantly from Kant's actual views about mathematics. This comes out most clearly when he paraphrases Russell's characterisation of the Kantian

[2] So that even Russell was later to adopt a view on which numbers are classes, and classes are merely logical fictions.

position on geometry.[3] 'If arithmetic has apodeictic certainty, its matter, i.e. number, must be *a priori* and as such must be purely subjective; and conversely, if number is purely subjective, arithmetic must have apodeictic certainty' (CA p.174). Russell states the Kantian position on geometry only to question it; Bachelard paraphrases Russell in order to affirm a position on arithmetic which is in some respects Kantian: 'Number is only a moment of numeration and all numeration is a method of thought. One could also say that number is a synthesis of acts' (CA p.174). Numbers and arithmetic have a wholly subjective origin and correspond to an internal experience: the experience of a rationally active subject. The conditions of rigour are, Bachelard says, inextricably linked to those of voluntary action. 'If we want to know with a maximum of rigour, we must organise acts, substitute totally the constructed for the given' (CA p.174). The bare structure of constructive thought is that of the step-by-step 'construction' of a number in counting.

Here Bachelard is, however, thinking in terms of rational, intellectual construction, not of construction of images in the imagination or in *a priori* intuition. Construction as a rational process, which can yield knowledge in so far as one has a conception of and hence knows the method of construction, is a discrete and linearly ordered process, one whose formal archetype is to be found in the construction of the natural number sequence. Grasp of a *method* of construction does not carry with it an ability to visualise or imagine the product. Taken by itself it yields only an intellectual, non-experiental conception of the entity constructed.

In this respect Bachelard departs further from Kant than does Brouwer. Although he would agree with Brouwer's conception of mathematics as inner architecture (Brouwer 1964a p.84), there is a subtle difference of emphasis between Brouwer's basic intuition of a 'bare two–oneness', derived from 'the falling apart of moments of life into qualitatively different parts, to be re-united only while remaining separated by

[3] This is given in Russell 1897 p.1.

time' (1964b p.69), which he considers to be the fundamental phenomenon of the human intellect, and Bachelard's experience of number as a synthesis of acts. In the former the subject is still essentially passive; the intuition is derived from a reflection on passive experience. In the latter the subject is essentially active, reflecting on his own activity.

Moreover, Bachelard is not nearly so hostile to formalisation as either Brouwer or Poincaré. Indeed, he takes Poincaré to task for complaining that he could make no sense of Peano's formal axiomatisation of arithmetic. This, Bachelard says (NES p.35), was because Poincaré treated Peano's notation merely as an isolated and conventionally established vocabulary without really trying to use it. For one only has to apply Peano's formulae to feel that they reflect our thought about numbers, that they regularise it in such a way as to leave us wondering about the source of our feeling that they do capture entailments (about the origin of the force of the arithmetical 'must'). The dialectical play of form and matter is more deeply embedded in our thought than we might think. It is necessary already to have experienced mathematical thought at the unreflective, common-sense level before it is possible to construct an axiom system in such a way that the axioms command respect. The standards of correct reasoning appear then to be located in the formal rules, in the conventions for the use of formal symbols, and yet they can only be recognised as standards to be obeyed in reasoning if they regulate an antecedent practice, and hence only on the basis of a prior participation in that practice. It is this which gives the sense of an interplay of form and matter. 'Hereafter an axiomatisation *accompanies* scientific development. The accompaniment has been written after the melody, but the modern mathematician plays with two hands' (NES p.36).

4 TWO-HANDED IMPROVISATION ON A THEME

It is just this two-sided nature of modern non-Euclidean mathematics, the formal axiomatic accompanying the informal, which Bachelard sees as a new and characteristic feature. There is a constant demand for formal rigour, but this is always for the rigourisation of a non-formal practice and without engagement in this pre-existing practice, the formal axiomatic system which rigourises, rectifies and extends it will make no sense, will carry no psychological weight as a system of principles of reasoning, of principles which *ought* to be followed.

a *Left hand*

Bachelard sees the demand for rigour as initially arising out of the recognition of error. Errors occur when we try to use intuitively based concepts and methods in domains other than those which gave rise to them, as when we try to make use of geometrical intuition as the basis for reasoning about infinitely small quantities. The derivation of paradoxes or contradictions (such as those of naive set theory) prompts a re-examination of the concepts and methods which had been taken for granted. It is only when problems occur that we are prompted to reflect on what we have been doing and, in a move toward abstraction, to 'disengage the method from the conditions of its employment' (CA p.173). It is when we try to extend the application of concepts and methods beyond their original domain that we are forced to reformulate our concepts and 'abstract them methodically from the intuition which proposed them to us' (CA p.169). It is this abstraction of procedural form from intuitively grasped content that places mathematics in the domain of reason.

Mathematics starts in experience, from what is given there. But the ideas of proof and of a mathematical object as something about which things can be known by means of proofs are not given in experience. It is only to the extent that mathematical objects are constructs that certain knowledge of their properties is possible. But what sort of constructs? What is construction here? It is not, as with the piece of wax (pp.36–8), methodical physical preparation. Nor can it be purely mental construction, for mental constructs are problematic indeed as objects of public, objective knowledge. Rather, the construction goes on in the development of organised mathematical theories; in the process of agreeing on precise definitions finding adequate sets of axioms, and standardising proof procedures. In other words, it takes place in the process of formalisation, of imposing standards of rigour. As in the case of the piece of wax, it is not a matter of free, unconstrained construction, for it is a methodical *re*construction of something which was given. But nor is it a fully determined construction; there are decisions, choices to be made. The direction of future development is not determined although it is constrained.

One way of elaborating Bachelard's position might be as follows: Shapes (circles, squares, triangles, etc.) are perceptually distinguished and recognised independently of any knowledge of geometry and its definitions. As such they are not geometrical objects – objects of geometrical theory. The problem of finding definitions of such shapes is set, on the one hand, by the kind of questions that a theory of shapes, or of spatial configurations, is required to answer, the sort of knowledge of these properties that is required, and, on the other hand, by the methods available for answering such questions (where the methods may themselves throw up some of the questions). Where there are questions which can be intersubjectively posed, there is room for counting their answers as some kind of cognitive advance, if one has reason for believing them to be correct. Definitions will be 'correct' to the extent that they are in approximate accord with the perceptual concepts they purport to define, and they will be adequate to the extent that they allow some of the questions to be answered by the methods available. The provision of postulates rigourises and serves as basis for the legitimation of methods of proof. The postulates themselves need to be pre-theoretically self-evident.

Greek geometry, as a theoretical discipline, reveals its origins in non-theoretical practices of working out practical problems of land measurement, architectural construction, astronomical observation, etc. – practices which were based on diagrammatic representations, drawn using straight edge and compass. Geometry, as a body of abstract theory, could be seen as arising in response to questions about these practices (reasoning practices which already involve the unreflective use of representations) in much the same way that syllogistic theory can be seen as arising in response to questions about the argumentative practices used in debate: questions about how to distinguish between legitimate and illegitimate reasoning. Such questions ask for standards to be explicitly formulated, for the practices to be codified. But because they are questions about *reasoning* practices, they are not self-contained, standards cannot be arbitrarily set. They are constrained by the requirement that the reasoning endorsed as legitimate be reliable. Such questions thus require, for their answers, reflection on the possible grounding of the distinction between correct and incorrect moves – a theory.

This would be one way of interpreting Bachelard's notion of mathematical abstraction as a matter of 'disengaging a method from its conditions of employment' (see p.82). It would see geometrical theory and its objects as arising not out of direct abstraction from sense impressions of empirical circles etc. but out of the attempt to conceptualise and systematise a pre-existing body of measuring and calculating techniques, ones which already employ representations (constructs). It is reflection on rational *activity* which forms the basis for abstraction. Geometrical objects would then be abstract, idealised *re*constructions of diagrammatic representations, with their definitions derived from reflection on items which, in already being constructs, can be known as such by reflection on and conceptualisation of methods of construction, when the methods are isolated from the practical, physical conditions of their employment.[4]

But once Euclidean geometry is axiomatised, with its definitions and axioms accepted as intuitively correct, what

[4] This interpretation makes Bachelard's notion of abstraction very much like Piaget's notion of reflective abstraction. See Piaget 1966.

further development can there be? If mathematical theorising ultimately arises out of the systematisation of pre-existing reasoning practices, one can understand how it is that these theories should appear to have content, a subject-matter of their own. There are pre-existing standards against which attempts at systematisation are checked. If this were the only kind of cognitive step made in mathematics, one could easily see how the traditional realist view of geometry and arithmetic arose. Once Euclid's axiomatisation had been achieved and accepted, it provided a fixed framework with a determinate domain of objects (a well-defined domain of intended interpretation) and a determinate body of possible knowledge of them (the set of necessary truths entailed by the axioms and definitions).

However, the successful application of non-Euclidean geometry suggested that not only are geometrical objects the product of reflection on our own representational practices, but that there is nothing unique about these practices. As we can also use non-Euclidean geometries as the basis of our representations, it appears that we are free to choose the forms of our representations to suit the situation. In this case, surely, the forms of representation can be completely abstracted from the representational practices which gave rise to them. It was already clear within Euclidean geometry that systematisation makes it possible to extend our powers of representation and reasoning well beyond their initial intuitive base. Figures could be defined and theorems proved about them prior to any diagrammatic representation and prior to encountering any instances in experience.[5] Can not the same now be recognised to be the case for whole systems, so that they can be defined and results proved in them prior to and independently of any application?

In one sense this is true. Formal logic formalises and systematises the process of axiomatisation itself in such a way that one can now have a mathematical study of formal systems, a theory in which formal systems are themselves the object of

[5] Cf. Descartes' discussion of the chilliagon at the beginning of *Meditations* VI.

study. All sorts of formal systems can be constructed and studied in this way. From the purely formal point of view the mathematician is completely free; unconstrained by the empirical, mathematics can become an exercise in free creation. From this point of view any consistent set of axioms may be chosen and taken as implicitly defining a set of terms, whilst at the same time bringing into being the objects of which the axioms are 'true'. The definitions (axioms) create the objects; the axioms are their entire being. There is no prior intuition constraining such constructions. Such mathematical objects would be wholly constituted by the relations established by freely chosen rules (CA pp.179–84). But of objects so defined Bachelard says 'they have not the least fertility' (CA p.179). The price of freedom is that free creation cannot at the same time claim to be a source of objective knowledge. The production of a formal theory may be a creative act, but it does not in itself represent a cognitive advance. If there is no more to mathematical being than we arbitrarily define into it, there is no cognitive gap, nothing to be 'discovered', nothing unknown.

This is not, of course, literally true, since the number of theorems of any formal system is potentially infinite; no one can claim to know all of them. Yet once the system is defined, the set of theorems is also in a sense fixed and determined in that what can constitute a proof of a theorem is fixed in advance. Formal systems are in this respect closed rational systems, and indeed it is just this closed character that makes them possible objects of mathematical investigation. Objects which have being only within such a system are knowable in one way only; there is no alternative means of epistemological access to them. For example, numbers as given and defined by formal number theory are presented in an ideally neutral and rational reference frame – a purely formal one. In this sense they have achieved independence of the individual subject and his subjective modes of knowledge. The criteria of proof are wholly intersubjective and indeed are such as to make the checking of proofs a mechanical process. Yet, once divorced from the subject in this way, they as it were lose a dimension and, as ideal objects identifiable only within an ideal framework, remain only ideal objects, objects fully given by the formal framework. If there are statements about them which are undecidable within that framework then those statements cannot be thought to be either true or false. It is in this sense that there is nothing to be discovered, nothing concealed from view, because there is no *other* point from which to view them.

Many people have indeed located the distinction between mathematical and physical objects just here. Mathematical objects lack the multi-dimensionality of physical objects. There is always another direction from which to approach a physical object; the criteria by which we identify and individuate such objects neither exhaust their properties nor serve as a basis for deducing them all. (In Locke's terminology, currently enjoying a revival: whereas physical objects have both a real and a nominal essence, mathematical objects have only a nominal essence.) If mathematical objects are defined into being it would seem that we are free to introduce any definitions we want so long as they do not contain contradictions. In this case the conclusion to be drawn from reflection on the development of non-Euclidean geometry is that there is no objective knowledge in mathematics; mathematics merely provides us with a multiplicity of symbol systems, of languages.

b *Right hand*

The formal accompaniment is in itself neither the source of mathematical innovation nor of the substance of mathematical theories. It leads only to closed and empty rational structures (to tuneless rhythms). Mathematical thought is not, Bachelard thinks, spurred only by the need, in the face of contradictions or paradoxes, for formal rigour. It is also driven by a desire to unify, to extend the application of theories and techniques beyond the bounds originally set for them. Such extensions are not explicitly catered for by any systematisation of the original domain. They can, therefore, only be made informally at the beginning, and will continue to be made informally until such time as problems arise and the demands for rigorous systematisation once more become urgent. This second dynamical factor ensures that the task of formal systematisation will never be complete. But more than that, it is in the attempt to extend methods beyond their originally intended range that Bachelard sees the emergence of an analogue to the 'multi-dimensionality' possessed by physical objects, an analogue which grounds a metaphorical application of the concept of objective knowledge and of the possibility of progress towards it within pure mathematics.

What is it to believe in the reality of something? Bachelard's answer (NES p.34) is that it is essentially to be convinced that there is more to it than is evident from what is immediately given, to believe that there is objective knowledge to be gained. But how can this conviction arise in the case of mathematical entities such as numbers? This Bachelard admits is one of the most difficult and delicate cases, and yet just for that reason it is the most instructive, for it is mathematics which is seen as mediating the transition from subjective to objective in scientific thought. Mathematical thought has its origins in the subject, in reflection on his own rational activities, but as in the case of alchemical and pre-scientific thought, there is a tendency to objectify and project the characteristics of our own thought as features of an independent reality. The mathematician shares with the poet the tendency to ontologise, to give reality to his creations (NES p.35). We need to resist the temptation to realism in mathematics if we are to be able to discern the input on the part of the subject to scientific thought. But we also need to locate the origin of the temptation if we are to understand the nature of mathematical thought and hence the role that it plays in science.

What, then, gives rise to the impresson that mathematicians are explorers of uncharted realms, are discoverers and not creators? It is, Bachelard suggests, the fact that when we try to extend methods from one domain to another, to unify and generalise, we meet resistance. First attempts frequently fail, resulting in contradictions and paradoxes. 'This heterogeneity of domains is the source of resistance to the reciprocal assimilation of mathematical notions. It is this which, we believe, gives existence to rational entities' (CA p.188). This 'existence', he says, arises from the relative opacity of two different methods of reasoning. This will explain the temptation to realism in mathematics only if it is the case that in mathematics we are, at the outset, confronted by heterogeneous domains, by a lack of unity in mathematical thought. This is, Bachelard thinks, the case, and the fundamental division is that between the mathematics of the discrete and of the continuous. It is this division which he sees as playing within mathematics a role analogous to that between theory and experience in empirical science, and which therefore gives to mathematical thought more than one dimension. It is in trying to measure the continuous by the discrete that a mathematical reality seems to emerge, for the continuous resists analysis by the discrete, and this resistance is taken as a sign of independent existence. The irrationality of the continuum is taken as a sign of its independent reality (CA p.177).

This 'quasi-reality' of mathematical objects emerges extrinsically. As mere products of definition they have no independent reality. They take on this character only in relation to other mathematical systems, when there is an attempt to model one mathematical theory in another, or to apply one mathematical theory to the domain of objects defined within another, i.e. when heterogeneous domains 'interfere'. If this kind of cross-application is always possible, then no mathematical theory can form an entirely closed deductive system, for the relation of the objects of one mathematical theory to those of another cannot be fully determined in advance within either theory. They cannot be incorporated within the postulates of a theory since they may well depend on future mathematical developments, developments which are not predictable. It is the idea that there will be future development that gives the impression that the nature of mathematical objects is not fully captured even in their most recent definitions, that their explicitly systematised interrelations do not exhaust those which are implicit in their nature. But this is the element of illusion, for it is not what is now determined which forms the sole basis for future development; it is that future development itself, one in which there is an element of free choice, which will add dimensions and so alter our perspective on our present mathematical methods and the entities which play a role in them. In knowing that the general tendency of development in mathematics is towards abstraction, generalisation and unification, we have a sense that there is a yet-to-be-given more general classification of the kind of structures with which we are at present concerned. But there is no truly objective classification of created entities, for future classification will also be of our own making. Hence 'the reference of mathematical entities to the most general notions which will play the role of species and which will provide a foundation for their reality, is illusory' (CA 179).

Here we have a very abstract characterisation of the epistemology of mathematics. We can try to see what it amounts to by looking further in the direction in which Bachelard has pointed us, namely at the history of the successive attempts to unify theories of discrete and continuous magnitudes. Here four things will emerge: (1) that the second reflective abstraction (mentioned on p.85) prompted by non-Euclidean geometry was premature and based on an oversimplification of the situation; it ignores the structure of the developments which took place within Euclidean geometry itself between the time of Euclid and that of the discovery of non-Euclidean geometry; (2) the structure of some of those developments; (3) the sense in which Bachelard insists that mathematics is not just a language but is a rational framework, a way of thinking; and hence (4) the sense in which the structure of mathematical development is also the structure of the development of the rational, theoretical framework of scientific thought.

5 TO MEASURE THE CONTINUOUS

a *To rationalise the irrational*

The lack of unity which ensures that mathematics has a potential for development toward unity has been present from the time of Greek antiquity; the scandal of the irrational is the mark of the fundamental nature of the division and of its recognition.[6] This draws reflective thought up sharply. The presumption, inherent in practices of measurement, that the application of numerical concepts and techniques can be extended beyond the domain of the discrete to that of the continuous – transferred from counting sheep, thereby estimating the size of a flock, to estimating the size of a field or the length of a road by counting off units – is shown to be questionable, and indeed to stand in need of questioning. It presumes that there is a unitary and coherent concept of size embodied in the set of techniques used.

Yet the move of making such an extension of the use of numbers is merely a continuation of the process which introduces *numbers*, as distinct from things numbered, as not merely concepts with a general application but as entities to which abstract operations can be applied and whose results hold good no matter what is being numbered.[7] At the pragmatic level of measurement, the extension is wholly natural. Yet it is an extension which raises practical problems of finding units, of measuring areas of figures of a variety of shapes, etc. These are problems which require theoretical solutions. However, it is only when arithmetic and geometry become independently abstract that the extension presents itself as a problem in the form in which Bachelard characterises it, for it is only then that one can talk of numbers and figures as mathematical objects.

Arithmetic and geometry appear as independent disciplines, employing different methods and concepts. Whereas arithme-

[6] Although it is not entirely clear just how scandalous the Greeks did find this. See Fowler 1979.

[7] Cf. Goody 1977. He describes the numerical practices of the LoDagaa, who apparently can make no sense of the idea of just counting. One has to know what kind of thing is to be counted, for different kinds of things are counted in different ways, since the procedures involved require physical manipulation and cows and cowrie shells cannot be handled in the same way.

tic is concerned with calculation, with inferences relating to the precise sizes of things, geometry abstracts from size and deals rather with shape, with spatial relations and with the establishment of relations of size comparison between lengths, areas, volumes and angles. But what is remarkable is the appearance of a deep connection between shape and number (not shape and size) via the notion of proportion (or ratio), which is itself the basis of the measurement of continuous magnitudes. Geometrical objects, defined in terms of idealised ruler-and-compass constructions, seem to have inherent numerical properties, for these definitions not only fix the shape of a figure but in doing so fix the various proportions of its parts and the proportional relationships which it bears to other figures whose construction is linked to the one in question. It is the discovery of those proportions and of the constructions which realise given proportions which formed the concern of classical Euclidean geometry. For in practical terms, the discovery of proportionality relations is what is important for the purposes of measurement.

Discovery? This seems the only way to describe the epistemological situation. The arbitrary step is in focussing on ruler-and-compass constructions (it is conceivable that, if pre-theoretical practices had been different, different starting points might have been chosen). Even given the nature of the basic constructive operations, to know what it is possible to achieve by means of them is not a matter of pure logical deduction from Euclid's axioms. It is something which has to be quasi-empirically discovered. Possibilities have to be demonstrated by showing that they can be carried out, where the standards of demonstration are set by recognised proof procedures, procedures which rely on an intuitive grasp of the nature of geometrical construction and of relations of quantitative comparison between extended magnitudes.

It is inherent in the notion of a geometrical construction that figures produced by the same construction should be the same (similar, but not necessarily congruent). This is the rationale for thinking that constructive definitions can be given for shape concepts. Further, similar figures have the same proportions – the relations between their corresponding parts are the same. Thus it must follow from the way in which geometrical objects are defined that proportions are fixed by their definitions. This is to say that one can deduce that geometrical objects are inherently such that numerical concepts can be applied to them via the notions of proportion, ratio and measure.

But this theoretical fixing of the idea that it makes sense to talk of *the* ratio of the diameter of a circle to its circumference does not fix the means of determining what, numerically speaking, that ratio is, nor could it, for numbers, in their use as measures, do not enter into geometrical definitions or axioms. Here we have a situation in which a ratio is located and identified within one theoretical framework without thereby being known *as* a numerical ratio, without thereby being located within a system of ratios. It is this apparently justified postulation of a ratio as identifiable within two frameworks, whilst actually being identified only in one, that creates the appearance of a cognitive gap, of something to be discovered. The geometrical determination of ratios requires a construction permitting intermediate comparisons. So the application of numerical concepts, and the numerical determination of ratios, depends essentially on determining what constructions are possible. But the methods for turning purely geometrical constructions into ways of making numerical determinations of ratios between lengths, areas or angles, themselves have to be created and argued for, for they are given neither by the arithmetical concepts alone, nor by the geometrical concepts alone, although they are constrained by both.

Measurement practices have already fixed the way in which number is to be applied in the comparison of continuous magnitudes. In particular there is the constraint that inequalities must be preserved; if the area of A is greater than that of B (e.g. because B is contained in A) then the number assigned to A as a measure of its area must be greater than that assigned to B. It is this which forces recognition that there is a problem about numerically determining the ratio between the diameter and the circumference of a circle. One cannot arbitrarily fix on a rational number as giving the ratio because this would conflict with the conventions already recognised. One can generate (method of exhaustion) series of inscribed and circumscribed polygons, of increasing numbers of sides which put tighter and tighter bounds on the value, but which have to be recognised as not giving the value itself. They can only give a sequence of approximations of the form $p/q < \pi < p'/q'$. Comparison of spatial magnitudes is already geometrically constrained independently of numerical determination, just as numerical relations are constrained independently of the application of numbers to geometrical magnitudes. Even within the system of numbers itself, the question of algebraic extension arises. The positive square root of two can be referred to; one seems justified in presuming that there should be such a number, particularly as the ratio $\sqrt{2}:1$ can be realised geometrically. And yet it can be proved that $\sqrt{2}$ cannot be identified with any ratio of integers.

How exactly should the situation be characterised? We have two systems of concepts which, if considered abstractly (i.e. not genetically through their practical development), can be introduced axiomatically and hence as systems of conventions. These systems, formally regarded, are self-contained.[8] What counts as a proof is given within the system (as conventionally established) and the kinds of things for which proof can be expected are delimited by the language in which the axioms are expressed. One could only expect to be able, logically, to prove relations between the concepts occurring in the axioms or between ones explicitly definable in their terms.

But numerical concepts are developed to be applied, and if the axioms may be seen as expressing conditions of the possibility of their application, they do not determine how they are to be applied. Geometrical-magnitude comparison seems to legitimate the application of numerical concepts as measures of

[8] From this formal point of view one can, as in measure theory, consider the whole situation much more abstractly in terms of the general problem of the imposition of a measure (or of a metric).

those magnitudes. But with such application we are led to ask questions phrased in a mixed language of arithmetic and geometry, we ask numerical questions about ratios between geometrical magnitudes. Asking for a numerical determination of the ratio between the circumference and the diameter of a circle can be reduced to a purely geometrical question – asking for a method of constructing a square equal in area to a given circle. Geometrically, the introduction of the ratio seems legitimate, but then an alien notion is introduced – knowing what the ratio is, by which is meant knowing what numerical ratio it coincides with. The procedures for such a determination are not given within either system, but have to be created, and yet they cannot be freely created, for there is the constraint of keeping consistency with both systems. For without that we would not be answering the original question which demanded a relation of concepts given by those systems.

But equally, once numerical questions are asked about geometrical relations, one is led also to ask new questions about numbers and their relations. In particular, greater importance is given to inequalities (as opposed to equalities) and their manipulation, to multiplicative structure rather than to additive structure, and there is pressure to extend the system to include fractions as numbers. In this way cognitive gaps appear in both systems. Yet if we are, in the end, dealing only with conventionally instituted concepts, the questions introducing this depth, giving the appearance of additional dimensions to numbers and figures, do not necessarily have predetermined answers (the truth of the matter may not be uniquely determined). There are constraints on the ways in which we can try to interlock the two systems – consistency constraints – and these give an appearance of objectivity. But we have to discover not only answers to the questions, but also ways of answering them; we have to create a framework for doing so, and there is no unique way of doing this.

b *To algebraise geometry*

In this, as in many other areas, medieval thought gives no manifestation of the drive for a unifying synthesis. It treated discrete and continuous magnitudes separately, arithmetic extended by an algebraic theory of non-continuous proportions dealing with discrete magnitudes, and geometry dealing with the continuous magnitudes.[9] But what did happen was that geometry itself, as the science of continuous magnitudes, came to be used as a rational framework within which to represent and think about non-spatial magnitudes, such as weight, speed and time, and their interrelations. So that there grew up a *non-formal* body of techniques for using geometrical representations in this way.[10] Thus Descartes explains that by dimension he understands 'nothing but the mode and aspect according to which a subject is considered to be measurable' (Descartes 1931 p.61), so that weight, speed and many other things besides length, breadth and depth are dimensions. He then goes on to say:

> Recognition of this fact throws much light on Geometry, since in that science almost everyone goes wrong in conceiving that quantity has three species, the line, the superficies, and the solid.... And though these three dimensions have a real basis in every extended object quâ extended, we have nevertheless no special concern in this science with them more than with countless others....
>
> (Descartes 1931 p.62)

However, this extended use of geometry also created a pressure to extend the bounds of traditional geometry – to extend the domain of geometrical objects beyond those constructible by straight edge and compass to include at least the so-called mechanical curves, amongst which the conic sections were classified. It was Descartes' view that:

> if we consider geometry as a science that teaches a general knowledge of the measures of all bodies, we must no more exclude complex lines

[9] The same lack of demand for rational unification is apparent in the encyclopaedic character of many medieval treatises. The Greek demand for rational unity, particularly evident in Plato's works, might thus be argued to be one of the cognitive standards revived during the renaissance.

[10] For example, see Galileo 1638.

from it than simple ones, provided that we can conceive them as being described by a continuous movement, or by several successive movements of which the later are completely determined by those which precede: for by this means we can always have an exact knowledge of their measure.

(Descartes 1965 p.191)

He goes on to achieve a more systematic extension of geometry by *re*formulating it in an algebraic guise. The objects of Cartesian geometry are those for which a polynomial equation can be written (Descartes 1965 p.193). This not only has the effect of extending the class of geometrical objects, but also introduces a new kind of classification of them, one based on the degree of their defining equations.

These definitions are not arbitrarily introduced. To *re*formulate geometry, even with a view to extending it, requires that the new theory contain previous geometry within it; that the new definition of a circle, for example, defines the same thing as the old one. The locus, in a plane, of a point equidistant from another fixed point *must*, given the conventions for indexing points by rectangular co-ordinates, have an equation of the form $(x-a)^2 + (y-b)^2 = r^2$ (for a circle centre (a, b), radius r), as can be seen from the diagram below. Of course, if alternative conventions for assigning co-ordinates are used, the form of the equation will be different, so that in polar co-ordinates it would be $r^2 + a^2 - 2ra\cos(\theta-\alpha) = A^2$ (see diagram).

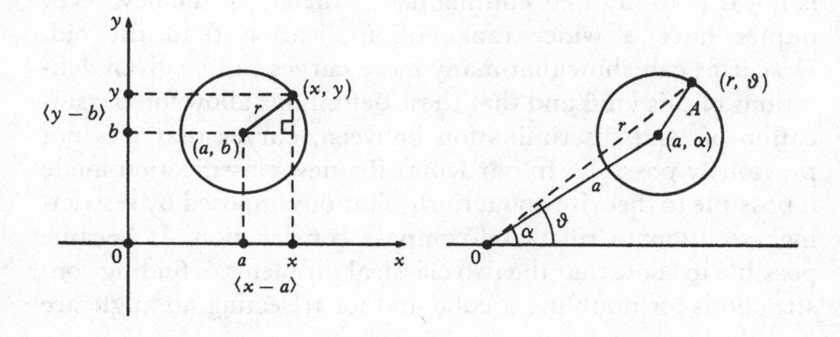

Indeed, once one has the conception of a co-ordinate system, geometrical theory itself determines how to translate from one such system into another, entails that there will be more than one such system, and determines what are the possible systems.

Descartes was interested in geometry primarily as a means for reasoning about the structure of relationships between physical magnitudes. His demands on geometrical definition were therefore different from Euclid's. A definition has to characterise the defined object sufficiently for demonstrations in the theory of proportions to be carried out; it has to make the object an object of an algebraised geometric theory. But at the same time it has to be *shown* that algebraic methods can legitimately be introduced into geometrical theorising. For this it is necessary to argue that old and new definitions amount to the same thing; to prove the correctness of the algebraic definitions. In other words, it is necessary to legitimise the introduction of equations as a new language which, where it covers the ground covered by classical geometry, is translatable back into the older terminology.

But linked to the new language will be new proof techniques, techniques directly applicable to the curves defined in this language, i.e. it is not just a language. In a sense one can say that nothing has changed, and one needs to be able to say this in order to justify new methods as methods of proof concerning the same (old) objects. It is for this reason that the intermingling of algebraic and geometric approaches is important. But it is not true to say that nothing has changed, for the new techniques have a wider range of application than the old. Descartes can show that many more curves can be given definitions of this kind and that these definitions allow for classification of, and discrimination between, curves that was not previously possible. In particular the new classification made it possible to theorise about the limitations imposed by restricting attention to ruler-and-compass construction. It became possible to show that the two classical problems of finding constructions for doubling a cube and for trisecting an angle are not soluble if the construction has to be by ruler-and-compass. For such constructions have the power only to solve quadratic equations, whereas the two problems reduce, respectively, to solving the cubic equations $x^3 - 2 = 0$ and $4x^3 - 3x - a = 0$, where a is a proper fraction.

So this is not just a change of language; it is a move to a larger rational framework within which to do geometry, one which does not leave the concepts as they were. Many more curves become objects of *mathematical* study, and thus subject to proofs (thereby becoming mathematical objects), but in addition, older methods themselves are characterisable and thus subject to analytic study. So this is not a conservative revision; it essentially changes the mathematics whilst including, in a new form, previous geometry, and it is important that it does so.

It is this which constrains mathematical development – there is a strong requirement of consistency with past, established mathematics, which is not merely that of a lack of conflict with it, by the development of something completely separate, but of the incorporation of established results. It is only this which entitles theoretical innovation to be counted as development or progress within a given mathematical domain, such as geometry, rather than the creation of a completely new domain. In the case of Descartes' geometrical innovations, one can see that the nature of this strong consistency constraint is to rule out certain moves as wrong, without the direction and actual form of development being entailed by pre-existing theory. It is a development which is rationally constrained by the two systems to be integrated, but which is not rationally required, and where the conceptual change involved in the redefinition of, for example, the circle can not only be rationally justified, but can be proved, within the wider framework, to be correct.[11]

[11] There is a similar pattern exhibited in the successive extensions of the number system. When the natural numbers are incorporated into the rational numbers, one has to show that 2/1 behaves as 2 should, and that arithmetical operations as defined for rational numbers give the 'right' results for the rational counterparts of the natural numbers.

But if mathematical truth is equated with this kind of correctness, it cannot be cashed out in terms of correspondence with an independently existing reality. In this sense the independent reality of mathematical objects is an illusion. The judgement of correctness here is relative to the frameworks involved, so that mathematical truth, like mathematical definition, is intimately linked to methods of proof, the means by which this 'truth' is established. More and more rigorous proofs can only be developed by the incorporation of a result into a richer system of concepts, a more general theory, with more powerful proof techniques. For this reason, Bachelard says 'Truth must be an accord of thought with itself; it is a property of knowledge which finds application at all levels of precision of that knowledge' (CA p.231). In other words, the account of mathematical truth here offered is a coherence account, where the requirement of coherence is the strong consistency requirement imposed in the context of recognition of rational unification as a goal. Its recognition requires acknowledgement of the possibility of redefining mathematical objects, making them susceptible to new, but as yet unenvisaged proof procedures. This simultaneously institutes a cognitive gap with respect to these objects, giving them the appearance of independent existents, and grounds the essential openness of the rational framework of mathematics.

c *To arithmetise analysis*

Descartes' first moves toward the algebraisation of geometry, far from instituting a new and settled framework, opened up a whole new set of questions. Most problematic was the underlying presupposition behind the move, the assumption that the Euclidean plane can be treated as a set of points each one of which can be indexed by a pair of numbers, co-ordinates, once an origin and pair of co-ordinate axes have been specified. Descartes could provide no justification for this assumption, for he had no theoretical account on which it was possible to understand how a continuum could be constructed out of discrete, extensionless points. The development of co-ordinate geometry and calculus together with their stunningly successful applications made more urgent the need to rationalise the continuum, to replace the geometrical continuum and reliance on geometrical intuition by a numerically based construct.

Central to this was the need to develop a numerically based theory of limits, incorporating a definition of 'limit' which would make clear the conditions under which limits of series exist and which would at the same time clarify the basis on which claims about them can be made. Given that there was a pre-existing geometrical notion of limit and of approximation to a limit it is natural to think of the project as one of finding *the* correct analysis of these notions. But it is not clear that there is, even here, a basis for introducing such a strongly realist conception of truth.

The definitions now accepted arose ultimately out of the introduction of calculus and the long sequel of attempts to make the infinity implicit in the continuum (in virtue of its infinite divisibility) amenable to rational comprehension by eliminating the need for infinitesimals. The justification for these definitions is to be found in the course of those developments. Success in the project was in part determined by reference to past accepted theory which constrains the definitions and accompanying reformulation of previous results. But we have no ground for supposing that there was only one way of achieving this. Indeed, Abraham Robinson's non-standard analysis has been interpreted by some as suggesting that a theory of infinitesimals could be (have been) developed.[12]

There is, moreover, a significant difference between the conception of a limit based on geometrical intuition and that provided within the arithmetical reconstruction of analysis. On the basis of geometrical intuition it seems clear that limits (points) have an existence independent of any sequence of interval approximations to them. But once numerical methods

[12] When one looks at Cantor's work there seems to be a certain degree of arbitrariness in his willingness to countenance infinitely large numbers and to make use of these in the analysis of the structure of the continuum, as contrasted with his total opposition to the infinitely small. See Dauben 1979 pp. 130–1.

can fully replace geometrical methods and reliance on geometrical intuition can be eliminated, the methods can be abstracted from the intuition which gave rise to them and one can then ask whether, whenever the numerical methods are applied, it is necessarily presupposed that when one talks of approximation to a limit one is committed to thinking of that limit as having an existence independent of the sequence of approximations to it. Bachelard argues that this is not the case, for the numerical reconstruction is achieved by defining transcendental and irrational numbers in terms of convergent sequences of rational numbers, i.e. in terms of sequences of which they are the limits. In other words, limits of series are not taken to have any existence independently of the possibility of knowledge of them (i.e. independently of possible routes of rational approximation to them) (CA p.211). The appearance of the independent reality of limits is retained by the fact that there can be two kinds of knowledge of them. There is the knowledge that a series has a limit or that an equation has a solution, and there is knowledge of what (number) this limit or solution is. For existence may be proved indirectly; if a series can be proved to be convergent, then it has thereby been shown to have a limit, but this proof does not necessarily identify the limit in question.

Thus when confronting (CA pp.232–3) the objection that all approximation necessarily presupposes a fixed number by reference to which the degree of approximation is measured, Bachelard responds by saying that there is no such presupposition. This is because (1) convergence is proved by inspection of a finite number of terms of a series together with the law which fixes the relation of consecutive terms, and (2) the degree of approximation is never itself exactly measured; it also can only be known by approximation. If we were to ask 'But how do we *know* that we are approaching something with our approximations?' there is a sense in which we would have to answer that we cannot *know* this. That the convergence of a series is proof of the existence of a limit is built into the definition of a limit and so is, to this extent, a matter of convention rather than of knowledge, even though the convention is well motivated.[13] Hence we have no ground for supposing that there has to be something, a limit, given independently of our definitions and to which a convergent sequence approximates.

But this is not taken to mean that we have no right to talk of limits as distinct from convergent sequences; Bachelard does not identify limits with either convergent sequences or with equivalence classes of such sequences (since there are, for example, a number of sequences which can be used for calculating approximate values of π).[14] What is meant by talk of limits, the sort of existence they have, is however to be understood only in relation to what counts as proof of their existence. The force of this commitment to limits as distinct from sequences approximating them is that it treats as admissible indirect or non-constructive existence proofs (proofs by *reductio ad absurdum*, which are not permitted by Intuitionists). It is at the same time admitted that this creates a gap between knowledge of existence and ability to identify, or have knowledge by approximation of the 'object' in question, and therefore it creates the appearance of a realist commitment, a 'metaphorical reality which is the sign of a heterogeneity between the domain in which the entity is posited and the domain in which it is studied' (CA p.195). In other words, it is just the sort of gap that arises as a result of the interference of domains, or

[13] This is built into Cantor's definition of real numbers in terms of (Cauchy) sequences of rationals. A sequence $\langle p_i \rangle$ of rational numbers *has a limit* (or *is convergent*) if and only if for any positive rational number δ there is a positive integer n such that the difference between p_n and any subsequent term of the sequence is less than δ, i.e. if $(\forall\delta)\ (\exists n)\ (\forall m)\ (|p_n - p_{n+m}| \leqslant \delta)$. The condition for two convergent sequences $\langle p_i \rangle$, $\langle q_i \rangle$ to have *the same limit* is then that there is some point in the sequences after which the difference between corresponding terms is arbitrarily small, i.e. $(\forall\delta)\ (\exists n)\ (\forall m)\ (|p_{n+m} - q_{n+m}| \leqslant \delta)$. This condition is given without making any reference to the limit itself. Real numbers are then defined by Cantor as equivalent classes of convergent sequences of rational numbers.

[14] Similarly he does not condemn the introduction of non-denumerably infinite classes to characterise the order structure of point continua, but rather insists that to know what can be meant by doing so one must go back to look at the way in which they were introduced, the kind of justification offered for doing so, and at the methods now accepted for proving assertions about such infinite classes. The introduction of non-denumerable infinities is an extension of a symbolism. If it is to be more than just a symbolism, it has to be possible to justify the extension of the number system by giving some sense to the symbols.

from the location of a (kind of) object in more than one domain, enabling identifying reference to be made to it from within two independent conceptual frameworks.

6 THE EPISTEMOLOGY OF REASON

This discussion of approximation is important, for the structure of the epistemological situation revealed in the reconstruction and replacement in analysis of the geometrical continuum by the real number continuum provides Bachelard with a model for some aspects of the epistemological situation in science. In particular it provides an example of how it might be possible to make sense of the idea that there is cognitive progress in science without having knowledge of, or even being committed to the existence of, any absolute standard by reference to which this is assessed, i.e. it suggests the possibility of a conception of increasingly objective knowledge which is not grounded in objectual realism or in a correspondence theory of truth. For Bachelard's general view of scientific knowledge is of a sequence of approximations, where successive approximations correct and reveal the errors of previous estimates.

Moreover, the mathematical notion of approximation arises out of attempts to comprehend and rationalise the continuum, to capture the continuous in the discrete. It is the product of the interference of independent domains, but of a particularly significant pair of domains, for Bachelard takes it that the intellect is discretely structured, because cognition takes place in a series of discrete acts, whereas experience has the structure of the continuum.[15] 'The discontinuous is clearly antecedent ... it is what casts light into this gloom. The continuous is thus only determined to the extent that the discontinuous determines it. In itself, it is not determinable' (CA p.221). Thus the interplay between the pure structures of the discrete and con-

[15] This is not a claim peculiar to Bachelard; it is, for example, a theme found, in slightly different forms, in both Bergson and Whitehead. It does, however, represent a marked contrast with the atomism of those British empiricists who, like Russell, took their cues from Hume.

tinuous magnitudes in mathematics also becomes the structure of the interplay between theory and experience in science. (See also CA p.175.)

The structure of this interplay resulting from the clash of heterogeneous domains (where attempts to extend the scope of one set of methods generate a need for theory, for abstraction of these methods from their initial intuitive base by means of a rigorous reformulation, an abstraction which itself results in further generalisation, suggesting further extensions in an unending cycle) is also realised by science in its theoretical endeavours. It incorporates as one of its standards of objectivity the drive towards ever more general, more abstract theories. This is the standard imposed by the conception of objective knowledge as that of a rational being. Mathematics, the paradigmatically rational discipline, exhibits the structure of development imposed by ideals of rationality in its purest form. But if mathematics develops and changes its character, the paradigm of reason also develops. Standards and forms of reasoning change but they do so in ways which are required to be rationally justifiable, even though there is an element of choice involved in the actual determination of the direction of the change.

An analogue for the model of rational development which emerges in Bachelard's work would not be that of a machine generating a recursively enumerable set (the set of theorems of a particular formal theory) but is more that of a person colouring in shapes made by a network of lines. The lines are given and constrain the shapes to be coloured, although not all need be heeded, since some may be coloured over and obliterated to make larger shapes. The first step may be random. Reflection on that choice may restrict the next. At each stage there emerge more constraints if one wants to be 'consistent' with past choices; there will be more involved in being consistent. It may be impossible to be consistent with all past choices, or there may be several choices which would be consistent with them all. That there is any notion of consistency here derives from the element of interpretation of past choices and the desire to see and continue any structure they embody, to see them as embodying principles. After routinely 'doing the same thing' for a number of steps, the result may reveal the possibility of a larger-scale pattern and of a new principle of repetition, repetition of a larger unit. Operation according to this principle may in turn reveal new and larger patterns. Such choices and their interpretation found subsequent choices, but do not determine them, although they do make mistakes very evident. (This sort of process might be thought of as a kind of recursion from the inside, rather than according to some externally determined principle.)

Putting the matter in this way serves to highlight the point which is central to Bachelard's epistemological approach, namely that this sort of rational development is not just a matter of the quasi-autonomous evolution of conceptual structures, but is an evolution in thought, or in ways of thinking. Most importantly it is an evolution which results from reflection on voluntary and intentional action. It is because, or to the extent that, mathematical objects are rational constructs (the products of intentional action freed from material constraints) that they are possible objects of demonstrative knowledge. It is because whatever principles govern them are principles of intentional action that they can be known by reflection. Mathematical thought is thus rooted in the subject, in the structure of his activities. What underpins the possibility of development in mathematics, development which involves the extension and modification of concepts, is the capacity of rational agents to reflect upon their own past actions, to ask questions about them, to seek to interpret them by seeking to make explicit the principles underlying the intentions with which they were performed, and to evaluate actions critically. But such reflection may itself be made the object of further reflective scrutiny; there is no end to the hierarchical structure of reflective thought. And this too is something discerned by higher-order reflection on the principles of rational thought and is a mathematically describable structure, as is the higher-order structure produced by the yet higher-order reflection. Mathematics itself exhibits and internalises the reflexive and open-ended structure of the thought of a rational, self-conscious subject.

The successive theoretical levels on which mathematical thought is conducted reflect levels of 'self-knowledge', knowledge of the formal structure of thought, of rational procedures. They also reflect degrees of integration of various ways of thinking, and therefore represent stages in the progress toward an integrated self-conception of the rational (reasoning) subject. The drive for unity, consistency and coherence is a requirement imposed by the rooting in a single (idealised) subject of various ways of thinking. But this is not the sort of self-knowledge to be gained by introspection; it is not empirical self-knowledge. It is theoretical knowledge of thought structures, of the structures of the thought of rational agents engaged in reasoning practices. And just as in the process of characterising its own proof procedures mathematics changes its own character, by adding a new level and introducing new proof procedures (as for example with Hilbert's metamathematics and the development of proof theory as a mathematical discipline), so too the critically reflective activity of rational subjects changes their nature, changes the character of rational thought.

What counts as rational thought has to be both intersubjectively and individually grounded; it has to be both explicitly formulated in formal, normative principles and implicitly, pre-formally experienced. The step repeatedly required is from participation in a structured practice to explicit formulation and knowledge of the principles which govern it, i.e. principles which, from an abstract point of view, can be proved to reproduce some aspects of that structure. The structure then appears as a construct, generated by formal principles, and so becomes an object of demonstrative knowledge. This is the constructivist core: the law, the principle of construction, is what gives rise to mathematical objects (cf. CA p.187). The successive levels at which laws operate and at which structure is perceived are necessarily accompanied by an open-ended hierarchically organised ontology of constructs.

But knowledge of such a construct is significant knowledge only to the extent that the principles generating it are recognised, by participants, as the principles governing, and thereby structuring, an actual practice. And that relation, although it may be a matter of intersubjective agreement or disagreement with the operation of intersubjectively imposed standards of correctness within the practice, is nonetheless essentially one which can be made only by a rational being capable of intentional action, capable of acting according to the conception of a principle. For it is only for such a being that the connection between actual action and its conception, between action theoretically conceived and reconstructed and action actually executed, can be made. It is thus the sense of self-identity of individual rational beings that confirms and underwrites the identity of subject-matter across conceptual revision, or redefinition. For redefinitions have to be recognised as such, they have to be seen as moves in the direction of increased rigour which incorporate at least some of the principles already implicitly employed and as making justifiable departures from past practice. Experience of the practices being rigourised is a necessary condition of the possibility of such recognition. If attention is paid only to conceptual systems, the identity underlying conceptual change, and hence the possibility of a rational grounding for such change, will remain hidden from view.

Here we see the force of the claim that even in the context of formal, axiomatic thought it is necessary to reintroduce a 'psychological' dimension and to run the gauntlet of the charge of psychologism. It is necessary to remember that the thought under discussion is the thought of someone, a subject, although (and this is crucially important) this thought is, in principle, the possible thought of any similarly placed rational subject. This is to be distinguished from the introduction of empirical psychology, or of a psychologism in which thought is treated as the thought of an empirical individual, thought which is private to him and to which he has, by introspection, privileged epistemological access. Indeed the focus is not on thought-*contents*, experiences or intuitions, but on thought-*activity*, its principles and structure.

7 THE STRUCTURE OF NON-EUCLIDEAN THOUGHT

The thesis, then, is that mathematics is not just a symbolism, a set of conventions for the use of special, formal vocabularies, but is intimately connected with the structure of rational thought, with reasoning practices. We have seen how, in the case of the introduction of co-ordinate geometry, geometrical concepts and methods of proof underwent a co-ordinated development. But this was not just a mathematical development; it provided a new rational framework for science. Perhaps most significantly it gave rise to the mathematical notion of a function and the corresponding scientific notion of a natural law.

Just as co-ordinate geometry inaugurated a new kind of scientific theorising, so too Bachelard sees the more recent changes in mathematics as having made possible further changes in the character of scientific thought. What, then, has been the effect of the development of non-Euclidean geometries? What kind of rational framework does mathematics now provide for science? One effect of the development of non-Euclidean geometries has been to alter completely the relation between a scientific theory which involves physical magnitudes presumed to be continuous and its mathematical framework. With only Euclidean geometry in the picture, there is no question about how to co-ordinate continuous magnitudes of different dimensions; they will be represented as dimensions of a Euclidean space. It seemed to past generations that, no matter whether one worked algebraically or geometrically, there was only one science of continuous magnitudes. But with the advent of non-Euclidean geometries there is a choice of spaces within which to represent and relate continuous magnitudes. This raises, for the first time, questions about the possible grounding of such choices; of whether it is just a choice of conventions, or whether it is a choice which can be empirically grounded.

From the mathematical point of view the effect has been to open up a whole new avenue of development and to serve to make very evident a pattern for the extension of mathematical thought. An axiom, once intuitively thought to be unquestionably true, is questioned and it is found to be possible to develop systems in which it is false and to give these systems application. In the process more general theories were born, ones which can provide overviews and can treat the relations between various geometries.

However, this development did not occur in isolation from the arithmetisation of analysis. The attempts to 'analyse' the structure of the continuum led to point-set topology, set theory, and the theory of transfinite numbers. Thus in the process of characterising the continuum, a whole body of theory concerning non-continuous but nevertheless rich and highly complex structures developed. This has the effect of proliferating still further the choice of representations for physical magnitudes. There is no need to expect that any magnitude which does not have the simple structure of the integers must be continuous. The whole history of arithmetisation, with the study of limits, of methods of approximation, coupled with topology and set theory, provides an enormously rich body of techniques for dealing with the discrete. (One important aspect of this is the development of sophisticated statistical and computational methods.)

The process of axiomatising geometry, even Euclidean geometry, from this discrete standpoint effects just as radical a rethinking about the nature of geometry and its role as did the development of non-Euclidean geometries. To produce a successful axiomatisation involves finding a sufficient number of conditions to impose on a set of points for Euclidean geometry to be possible, i.e. for the normal axioms to be true and for the notions, such as distance, which are usually taken for granted, to be applicable (to be defined over the set). This can be looked on either as an analytic or as a synthetic project. Analytic because the list of conditions represents an analysis of the intuitive spatial concepts which make Euclid's postulates seem self-evidently true, and synthetic because the imposition of these conditions represents a construction of a continuum out of points.

An analysis in the form of an axiomatisation has, necessarily, a generalising potential. In producing a set of axioms, the question always asked is whether any of the axioms on the list can be derived from the remainder, because if so it need not be on the list, and a smaller set of axioms can be used. If each axiom is independent of the rest, the axioms jointly embody a minimal set of conditions (although more than one such axiomatisation may be possible for the same structure). But in such a situation, it is always possible to consider a more general type of structure in which some of the axioms hold but not others. Indeed, models of structures in which one axiom fails are usually needed to prove the independence of that axiom from the rest, in the first place. There is thus here a potential for a process which on the one hand may lead to the development of increasingly general and more abstract theories, using more and more 'rarefied' concepts, with less specific content but wider application, and on the other, as a complementary counterpart, an increasingly detailed and complex theory of the possible structures falling under these general concepts. The necessarily double aspect of this kind of development, going in two opposite directions at once – towards generality and simplicity on the one hand, and towards particularity and complexity on the other – is a recurrent theme in Bachelard's characterisation of the structure of scientific thought.

Axiomatisation does not, however, always in fact lead to the development of a new general theory; it merely creates a situation in which such a development would be possible. For one thing, independence results are neither trivial nor easy to come by, as the history of the parallel postulate illustrates. For another, the models used to prove independence results may be too non-standard to be interesting, i.e. to throw any light on the nature of the theory being axiomatised. Another illustration is provided by the situation in set theory. The Zermelo–Fraenkel axiomatisation was presented by Zermelo in 1908. It was modified by Fraenkel in 1922 with a view to clarifying the status of the Axiom of Choice, which Zermelo had originally assumed as self-evidently true. But it was only in 1964, and then only by the introduction of an essentially new technique (forcing) that, with Cohen's proof (Cohen 1966), its independence from the remainder of the axioms was finally taken to be established along with that of the Continuum Hypothesis, of whose truth Cantor had been convinced, but which he had been unable to prove. However, the consequence of these independence results has not been the development of any new and more general theory, since basic set theory itself provides the framework within which exploration is made of the structures in which the Continuum Hypothesis and/or the Axiom of Choice hold and those in which they do not. Attempts have been made to relate the results of these investigations to other areas of mathematics, but the question of whether there are constraints which should entail either acceptance or rejection of either the Continuum Hypothesis or the Axiom of Choice is far from settled. So at the moment one could say that a spirit of non-Euclideanism reigns in set theory as well as in geometry, but that opinions are divided as to whether this must, in the long term, be accepted as the 'correct' attitude to take.

In the case of Hilbert's axiomatisation of geometry, the situation is somewhat different. The axioms can be shown to be relatively independent and they fall naturally into groups: I. Axioms of Connection, II. Axioms of Order, III. Axioms of Congruence, IV. Axiom of Parallels, V. Axioms of Continuity. The axioms of each of groups I–III deal with a particular kind of relation between points, or between individual points and those sets of points which are to be treated as lines. Of the axioms in each of these groups Bachelard says (CA pp.181–2) that they are entirely conventional; they determine what is to be meant by 'order', 'congruence' and 'connection' in this space by fixing the structure determined by each of these relations. It is only when they are combined together in a single system with the addition of axioms from groups IV and V that anything interesting emerges, because only then can the independently defined relational structures be made to 'interfere in a richer way and give rise to a genuine science'. This science is not a science of objects, but of *relations*, and this represents a new perspective on geometry. It is one which is also reflected in Russell's work as when he says (1919 p.59) that in geometry we are not concerned with the identity of points (as objects). We can introduce these as (logical) constructs in order to characterise the relational structure which interests us, but it is the relations which are of primary concern, not only in geometry but, in his view, in natural science generally.

This illustrates a shift within mathematics from thinking in terms of objects as primary and then dealing with their relations, to treating relations, relational structures and their interrelations directly. Structures, in their turn are characterised in terms of their interrelations, in terms of the groups of transformations under which they are invariant. 'One could say to a mathematical entity: tell me how you are transformed, I will tell you what you are (NES p.32).' The use of invariance conditions to characterise structures is (as we shall see in Chapter 5) of crucial importance to conceptions of objectivity in modern physics. The idea of a structure preserving mapping (a homomorphism, or simply a morphism) of one structure into another is a relatively new and very powerful idea.[16] The gradual shift of mathematics from being the science of magnitudes to the science of abstract structures means that mathematical science, once equated with quantitative science, expressed in the form of numerical laws, comes to be expressed in the language of structures (spaces), transformations, operators and invariance conditions. Thus Bachelard says: 'the idea of the composition of operations reunited in a group is, hereafter, the common basis of physical experience and of rational

[16] It is this idea which forms the basis of category theory and of attempts to argue that category theory provides a more adequate foundation for mathematics than set theory. Irrespective of the relative merits of the foundational claims, category theory and its widespread use in algebra provides a significant alternative, non-object-based approach and is a development which, had Bachelard known of it, would have strengthened his position.

research (NES p.38).' This shift in the way in which structures are characterised goes hand in hand with the reformulation of geometry in the manner of Hilbert and the formulation of other theories in the same style. It necessarily has an impact on mathematical physics, for it is, once again, not just a change of language, but is also a change of rational framework; it enables scientists to think about things in new ways (cf. ARPC p.42).

However, what Wittgenstein (1967 p.84) called the 'motley of mathematics' should not be forgotten: mathematics is not a theoretically unified discipline, but has different branches in which very different techniques are used and whose 'objects' of study are correspondingly different. The project of founding all mathematics on set theory tends to reverse once again the perspective on relations and objects, for in set theory objects take primacy. But not all of mathematics is actually done in set theory.[17] This merely serves to emphasise the fact that there is a wide range of choice; there are many rational frameworks within which the scientist may now daydream. This choice is not between languages, or between images, but between rational frameworks (systems of rationality), between methods of research (NES p.43). The choice of mathematical framework involves the whole problem of scientific knowledge of what is real. For once it is understood that experiment depends on a prior intellectual construction (a project), one is led to seek grounds for believing the concrete project to be feasible in the abstract bases of such constructions. The table of possibilities of experience is thus the table of axiomatic systems. 'One thus accedes to the physico-mathematical culture by reliving the birth of non-Euclidean geometry which was the first occasion of the diversification of axiomatic theories' (NES p.44).

8 LOGIC, MATHEMATICS AND SCIENTIFIC RATIONALITY

We now begin to get some sense of the consequences, for math-

[17] Exactly what should, philosophically, be made of the claim (substantiated at length by the Bourbakists) that nearly all of mathematics *could* be done in set theory is another question and one which cannot be pursued here. We can merely note that similar claims have been made on behalf of category theory.

ematical physics, of the divergence between Bachelard's conception of mathematics and that inspired by the formal foundational work of Frege, Hilbert and Russell. In this formal work it is presumed that mathematical reasoning is nothing more than logical reasoning – that any appearance of mathematical content can be located in the non-logical axioms of a formally axiomatised mathematical theory. This reduction of mathematical deduction to logical deduction seems frequently (albeit tacitly) to be taken by analytic philosophers of science to mean that mathematical deduction is merely abbreviated logical deduction, so that the use of mathematics in science is merely as a means of facilitating deduction; science could, in principle, get along without it.[18] If this is the case, then, from a philosophical point of view, there is no ultimate difference in formal kind between Aristotelian qualitative science, a non-mathematical biological theory and modern mathematical physics. The theories can in each case be looked on as consisting of sets of sentences, logically organised, where prediction and explanation are alike accomplished by logical deduction. The only difference is that contemporary theories, or contemporary philosophers in their analyses of theories, have at their disposal a logical framework wider than Aristotle's; most crucially it is one incorporating a logic of relations, making it capable of handling the relationships asserted by functionally expressed laws. On such a view, the basic epistemological situation in science is not in any way changed by the appearance of mathematics on the scientific theoretical scene. The role of mathematics could not, therefore, be crucial in any account of the epistemology of either the seventeenth- or twentieth-century scientific revolutions. The constancy, and thus the continuity, of the role of mathematics in science is underwritten by the thought that mathematics has to do with the form in which theories are expressed, but not with their content (cf. NES pp.55–7).

[18] In fact, the justification of this move is far from straightforward, and it is not even clear that it can be justified. See, for example, Field 1980.

If scientific theories are represented in this way as axiomatic systems, then theoretical change has to be modelled as a change from one such system to another. Further, if the provision of rational grounds for a theory change is taken to mean the provision of grounds from which to effect a formal logical deduction justifying the change, then it becomes clear that there can be no such grounds, for the theories concerned have been modelled as deductively closed systems, so that deduction will never lead outside the confines of a theory, or to its modification. Even Popper gives no role to deductive reasoning in the formation of a new theory or in its development out of pre-existing theories. Others, Feyerabend in particular, have gone further and have claimed that there can be no deductive, logical relations between theories on either side of a revolutionary change (in his terms, such theories are incommensurable and as such cannot even contradict one another – see Feyerabend 1975). This is so if theoretical terms are thought to be implicitly defined by theoretical postulates in which they occur rather than by reference to a theory-neutral observation language. (If all observation is theory-laden, different theories may impose different interpretations on a given set of observations so that there is assurance neither of a neutral observation language into which to translate all theories and so logically relate them, nor of a neutral observational base relative to which competing theories may be decisively tested.) In this case a change of theoretical postulates changes the meanings of theoretical terms so that it is impossible to see the statements of the two theories as being logically related at all. For example, this means that there could be no logical relation between statements of Newtonian mechanics and of relativity theory, so that a sentence expressing the conservation of mass has a different meaning depending on whether the mass involved is Newtonian or Einsteinian. What Einstein denies is not, therefore, what Newton asserts, and hence they do not contradict each other. If theories are treated as determining their own conceptual frameworks, a theory change involves also a conceptual change.

The whole difficulty here is that the Fregean logically foundational way of thinking, which separates justification from discovery, cannot model conceptual change. A Fregean concept cannot change; conceptual change can only be a matter of the replacement of one concept by another. This is quite explicit in Frege's work; he was hostile to the whole idea that the concept of number, for example, could be successively extended by algebraic closure (Frege 1959 pp.104–14). A Fregean concept is required to be such that it is determinate, for any object, whether it falls under the concept or not (Frege 1960 p.33), so that its extension cannot change. Thus, in the way in which Feyerabend poses it, the question of the justification of theory change is a question of the justification of a conceptual change which is seen as the replacement of one set of concepts by another. Obviously such a justification cannot be carried out wholly on the level of, or in terms of, the concepts in question.

From Bachelard's point of view this impasse is the result of trying to treat an epistemological problem, the problem of the way in which scientific knowledge develops and progresses, as a logical one. In so doing we (1) completely lose sight of the epistemological role of axiomatisation, and (2) become wholly insensitive to the changes which have taken place in mathematical physics in this century.

(1) In treating all theories on the model of axiomatic systems we can pay no attention to the relation between formal and informal theories and to the element of conceptual revision which goes on in the process of axiomatisation. Thus whereas Sneed, on the one hand, can say, 'the task of providing a logical reconstruction, as I conceive it, is one of codifying and systematising an existing scientific theory, applying no external standards of judgement beyond simple clarity and logical consistency' (Sneed 1971 p.4). Bachelard on the other hand sees axiomatisation as an attempt to regulate and to impose a rational order on practices which are recognised as lacking in rigour (and very probably in both clarity and consistency). This is necessarily an evaluative task. It is seen as part of the

dynamic process of the development of scientific thought. It has an epistemological function which, in addition, precludes the possibility of treating the resulting axiomatisations purely formalistically, for it is already essentially epistemologically engaged. Examples of such 'engaged' discussion of theory formulation would be the critical examinations of the foundations of classical mechanics to be found in Mach 1960 and Hertz 1956.

(2) The result of insisting that mathematics is not just a language, and of refusing the foundationalist move of trying to reduce mathematics to logic, instead seeing mathematics as providing rational frameworks for science, is to set science against a background of rational structures and rational methods which itself has a built-in dynamics. The rational framework of science is itself historically conditioned, for it changes with developments in mathematics. The rational framework for Cartesian science is not that of relativity theory, or even of nineteenth-century classical mechanics, and could not have been. This is one of the sources of discontinuity in scientific thought. Thought structured by contemporary mathematics, availing itself of the structures it provides, is different in kind from previous scientific thought and is not continuous with it. But this is not a discontinuity so radical as to preclude a comparison which situates past thought in relation to contemporary thought, for it is characteristic of the pattern of mathematical development that previous mathematical theories and ways of doing mathematics are incorporated within contemporary theorising, but in a new guise. Mathematical development is unlike scientific development in that it hardly ever involves the rejection of a mathematical theory which has once passed the initial hurdle of gaining general acceptance, although it does involve the reworking and correction of theories.

So one of the consequences of holding that mathematics is not just a language, not just a form of expression (a way of saying what could have been said without it), is the claim that contemporary scientific theories owe their existence, their possibility, to the development of the mathematical theories which were employed in their initial expression. In this vein it might be claimed that Newtonian gravitational theory was made possible only by the development of calculus, and that relativity theory was made possible only by the development of non-Euclidean geometries and tensor calculus. Indeed Bachelard says: 'Tensor calculus is truly the psychological framework of relativistic thought. It is a mathematical instrument which creates contemporary physical science as the microscope creates microbiology. There could be no new knowledge without mastery of this new mathematical instrument' (NES p.58). Against this background, problems of conceptual change take on a quite different form. The ways in which mathematics develops, the sense in which there are criteria for counting some innovation as a mathematical discovery, the way in which axioms are assessed and justified, all provide a basis for discussion of the rational justification of theoretical change in science. Mathematics deals only in the illusory or metaphorical reality of rational constructs, of abstract objects, whereas science is concerned with the concrete reality of objects first known through sense experience. The empirical, experimental dimension of science has to be interlocked with its rational, mathematical dimension.

4

NON-BACONIAN SCIENCE AND CONCEPTUAL CHANGE

It is characteristic of modern science to insist on the importance of detailed and precise experiments as well as on working toward very general abstract theories (see Chapter 1 p.10). Moreover, the striking feature of its experimental aspect is its degree of technological sophistication. This is another of the respects in which twentieth-century science differs from earlier science. It is to capture this aspect of modern science that Bachelard coins the term 'phenomeno-technique'; to advance from Baconian to non-Baconian science is to go from phenomenology to phenomeno-technique.

Just as in his use of the terms non-Cartesian and non-Euclidean, Bachelard described modern science as non-Baconian not simply in order to reject Baconian or inductivist accounts of it, but to stress that science does have an empirical base and that the nature of this base has changed. Science had to be Baconian before it could be non-Baconian; aspects of the everyday world, the world of common sense, have first to become objects of scientific investigation before they can be objects of scientific knowledge, and the kind of systematic, documented observation that was to be carried out by groups of scientists in New Atlantis is seen, by Bachelard, as being crucial to this process.

1 FROM COMMON SENSE TO BACONIAN SCIENCE

Bachelard sees the transition from common sense to science as involving a reversal of epistemological values, a devaluing of what seemed obvious (see Chapter 2 p.56). This process requires a turning away from objects as immediately given: 'In point of fact, scientific objectivity is only possible if one has first broken with the immediate object.... Far from marvelling at the object, objective thought must treat it ironically' (PF p.1). Objects of scientific knowledge have to be both empirically and theoretically known; they have to be, as Bachelard puts it, 'abstract–concrete' objects. But this is a characteristic which the objects of a scientific discourse acquire only progressively; it is a mark of success if its objects attain this status. Initially the empirical object, as an object of ostensive reference has to *become* an object of scientific investigation.

At first sight it might be thought that the object of ostensive reference is already fully 'objective' and does not depend for its identity on either the conceptual scheme employed by the pointer or on his beliefs about the object to which he points.[1] But on reflection it becomes clear that in so far as this is thought to be an object, something with an independent existence, its conception outruns the experience on the basis of which ostensive reference is made. Recognition of a substance by its immediately perceived qualities (white, powdery, etc.) may serve as a rudimentary classification, but does not satisfy even the interests of a cook, who, in identifying a substance as salt, flour or sugar, is projecting beyond the perceived qualities to dispositional, culinary properties. The criteria of recognition before use are one thing, the identity of the object recognised is, however, constituted in another, functional, culinary frame of reference. This does not take us beyond the world of common sense; even there the identification of something as an object, as a constituent of reality, is relative to some, however vague, conception of that reality. Ostensive reference succeeds only in a cultural setting (as Wittgenstein noted (1963 pp.1–30)). Everyday objects, as objects of common-sense knowledge, have their identity determined within a conceptual framework incorporating all the values of our culture.

These values are what the scientist, in his concern for objectivity, for knowledge of the object as it is in itself, seeks to

[1] Cf. Russell's referents of logically proper names (Russell 1972 pp.55–6) and Kripke's notion of rigid designation secured by ostensive baptism (Kripke 1972).

discard. But he cannot just discard them by discarding his everyday thought about an object, regarding it merely as that I-know-not-what to which ostensive reference is made, for that can form no starting point for investigation: 'the simply designated object is not strictly speaking a good rallying point for two minds attempting to *deepen* their understanding of the sensible world' (RA p.54). If two people want to engage in a joint investigation, they have to be able to understand each other and so agree about the object of their investigation. They must be agreed both about what phenomenon is under study and what they want to learn about it. This is something which they can only articulate progressively, via an organised experimental programme. Scientific investigation concerns phenomena which groups of scientists can jointly investigate and can discuss with assurance that they are talking about the same thing. There have to be agreed methods of investigation and of establishing further facts about such phenomena.

Moreover, the public identity of objects of scientific interest cannot rest wholly at the level of shared language. The objects of investigation cannot be just objects of discourse with their identity fully determined within that discourse (by the identity and individuation criteria embodied in the meanings of the terms referring to them). We look beyond the bounds of language when, for example, asking questions about the connection between the criteria used to recognise a substance, such as flour, and the account of what flour is. When we pose such questions and ask for a reason or a justification for using the criteria we do, we project the substance, flour, as something which is not fully known (whose 'true identity' is not fully given) in either perceptual or culinary frames of reference. The true identity of such a substance is what is to be discovered, so that it is initially thought of with question marks, as it were. The very way in which it is thought about or referred to leads outside language, for the very understanding of the way in which language is used to talk about the objects of scientific study involves grasping that there are questions to be asked about them, questions which can only be answered by looking beyond language. Thus terms designating such objects cannot be regarded as having a fully determinate meaning.

It is this question-raising thought of objects which marks off reflective, scientific thought from unreflective common-sense acceptance of things as they are. Yet it would be unduly paradoxical to say that the domain of objects of scientific investigation is wholly distinct from the domain of everyday objects. It would not only present us with two ontologies competing for the title of Physical Reality, it would also make the relation of science to the everyday world totally obscure.[2] However, this difficulty only arises if we put undue ontological stress on the notion of an object; what is true is that the concerns of everyday life, our everyday interests in the world around us, are not by and large those of the scientist (thought about the world can in this sense have different objects). What is meant can perhaps be illustrated by taking the electric light bulb as an example. (This is something which Bachelard gives as an example of what he means by an abstract–concrete object (RA pp.108–10).)

For the urban, late-twentieth-century child, light bulbs and electric light are a wholly normal part of the environment; they are not something one even really notices until there is a power cut. As such they are not objects of curiosity; they are not thought about in a questioning way. Obviously they go on when you switch the appropriate switch; obviously a light bulb when put in a light socket and switched on will provide light; that is what light bulbs are for, it is their nature. Their nature is from this point of view fully constituted by their function, their place in the environment. But just imagine Newton transported to the present day; imagine him confronted with the phenomenon. To him it would be utterly amazing, not just a marvel or a novelty, something to be gazed at in reverent wonderment, but because it is something he cannot comprehend since he cannot see how such a thing could be possible. He would see it in the context of his investigations of light and his theories about it, and in the wider context of his conception of

[2] This seems to be the problem which Gaukroger 1976 finds with Bachelard's position.

the possible causes of things. He would view it with a frame of mind used to asking questions and investigating things. Such an unfamiliar object would immediately be problematic for him; in thinking about it he would raise innumerable questions, ones which would already suggest lines of further investigation.

These questions could, of course, be answered for him if he were to engage a physicist in conversation who was able to fill him in on electro-magnetic theory. For the contemporary physicist the light bulb is not just a part of the environment which he takes for granted (although it will also be that for much of the time), but it is something of which he can give a theoretical account. For him it is an object locatable in a body of theory and hence is identifiable as a theoretical object. The theory explains how and why the light bulb works, gives laws governing electrical phenomena, and so on. It is this very body of theory which is responsible for the existence of light bulbs, in the sense that it makes them possible. It is from the standpoint of contemporary science that the light bulb is an abstract–concrete object; it is an object of theoretical scientific knowledge and at the same time is a part of the material world. For Newton, in the eighteenth century, it was not even a possible object of knowledge. There are not here three worlds, or three light bulbs. There is just one material item which is thought of, and hence constituted as an object of thought (and knowledge), in three different ways. An object of knowledge always bears the marks of thought. The object of scientific knowledge is thought of, and hence constituted, as objectively as possible; it is constituted by the network of its causal and theoretical relations, not by its relation to the subject. It is as an object to be located in Newton's framework of mechanical and optical theory that the light bulb is problematic and an object which raises certain kinds of questions.

Initial institution of material objects or natural phenomena as objects of scientific study can thus only proceed in the light of some vague, schematic (theoretical) conception of their nature. This directs the systematisation of empirical classificatory procedures, for to become objects of scientific investigation they cannot remain as isolated objects of ostension determined only by their relation to the pointing subject, or as isolated objects of wonderment. They need empirical location in a network of material and causal, not merely perceptual, relations, which make their empirical definitions both more precise and less subjective as the procedures are intersubjectively standardised. (It was this level of establishment as an object of study that N-rays (see p.60) failed to attain. Their classification as N-rays, however, already indicated the projection of the identity of what was perceived beyond what was (claimed to be) immediately given in experience.)

The idea that scientific enquiry proceeds first by delimiting its object by use of increasingly precise descriptions based on methodically organised empirical access to it is already found in Bacon's tables of induction (Bacon 1960), in Descartes' 'Rules for the Direction of Mind' (Descartes 1931), and Mill's Canons of Induction (Mill 1973). Bachelard is departing from the wholly empiricist picture of science by regarding this empirical classificatory and clarificatory stage as only a necessary first stage. He does not even claim that all sciences have progressed beyond this stage (nor need one assume that all can), but contemporary physics and chemistry, he would say, certainly have.

Putting the foregoing discussion in Bachelard's terms, one would say that the object of scientific knowledge has to be located in a problematic. The object is always an object of interest, one for which the process of objectivation has not been achieved, an object 'which does not refer purely and simply to past knowledge encrusted on a name' (RA p.55). As he says, the position of the scientific object, that which it is actually instructive to study empirically, is much more complex than either that of the immediate object of ostension or that of an object of discourse (the referent of a name), and it is much more 'engaged' in a cognitive sense. To grasp the object of knowledge one has to have some idea of the methods by which one can come to have more knowledge of it, and hence also of what it would be to have such knowledge (the form it would take, the questions it would answer). It is the surrounding problematic which supplies this, so the problematic is antecedent to all experience which is to be instructive. Such a problematic is initially founded on a specific doubt, a doubt specified by the object of knowledge. This questioning doubt will not at first be precise. It is part of the way in which progress is made that it becomes more precise; progress in part consists in giving more precise articulations to the questions to be answered and to the methods suitable for answering them. 'It is necessary to develop a sort of topology of the problematic' (RA p.56). It is in this process of making the problematic precise, of arriving at precise questions with a clear idea of what will count as answers to them, that the object is established as an object of public enquiry and public knowledge, for such articulation is a co-operative enterprise: 'to pose an intelligent question to intelligent beings is to bring about *the union of intelligences*' (RA p.56). The problematic already includes some theoretical elements (in a conception of the natures of things) and a methodology (an epistemology). It is these which have to be corrected, each in the light of results relating to the other. Both have to be brought into sharper focus.

2 APPROXIMATION AND THE SENSE OF 'REALITY'

Scientific progress is the product of a continual dialogue between reason and experience (pp.47–8), it is a product of the interference of theoretical and empirical conceptions of the items under investigations. But before going beyond the empirical, Baconian aspect of science to consider the nature of this dialogue, we need to give some further attention to its empirical voices, for it is in the nature of its empirical base that science is to be distinguished from mathematics; it is this which means that the epistemological structures resulting from the interference between empirical and theoretical conceptions of an object of scientific investigation differ from those resulting from the interference between discrete and continuous in mathematics.

The point of entry into empirical science from pure mathematics is provided by Bachelard's employment of the notion of approximation. By focussing on this notion we find both points of analogy and of disanalogy between the epistemology of mathematics and of natural science. The analogy is spelled out further in the conclusion to CA. The disanalogy between the two domains is brought out in Bachelard's contrast between mathematical induction and induction in the empirical sciences.

In the conclusion to CA Bachelard says that rectification *is* approximation, even though it seems that the term 'approximation' belongs to the language of the realist (since it seems to presume the existence of that to which approximation is made), whereas the term 'rectification' is part of the idealist's vocabulary (rectification takes place in pursuit of an ideal). But, he says, from the dynamic, epistemological point of view there is no difference here. To the extent that one seeks exactness, one also seeks an object, and vice versa. 'Whether one postulates a mathematical entity or one posits an object, the transcendence is the same since it prolongs the same endeavour' (CA p.295). What is important is that the object of scientific knowledge *is* transcendent, and is not immediately given in experience. The object relates to an ordered sequence of perceptions in the same way that a real number relates to an ordered sequence of rational numbers. The order here is given by increasing levels of precision – successive approximate determinations are increasingly precise. Again the idea of a pure substance and the associated progression of criteria of purity, or of time and the sequence of increasingly precise empirical time standards, would provide examples of what is meant.

The kind of physical realism involved is described (CA p.298) as a realism without substance; it is a form of realism which is to be distinguished from the objectual realism which was rejected on pp.42–4. Not only is substance not ontologically basic, but physical reality (the object of scientific knowledge) is presented as an entirely functional category of thought, one which has no descriptive content. This category comes into operation as one becomes convinced that the process of rectifying and increasing the precision of our knowledge of physical reality can be continued indefinitely, just as in mathematics, the notion of a limit comes into operation only in connection with sequences which one recognises as capable of indefinite continuation. But whereas in the mathematical case one can prove that a sequence of rational numbers is convergent, thereby proving the existence of a limit for the sequence, this is not true in the physical case. Even if successive theories T_n and T_{n+1} are related in such a way that the laws of T_n can be seen as first approximations[3] to the laws of T_{n+1}, we have no way of proving that this sequence will continue to converge.

In the mathematical case, proof is possible by a process closely related to complete induction. If the sequence is generated by a law (or can be approximated by a law-governed sequence) then in knowing the operation, repeated application of which generates the terms of the sequence, one knows in advance the relation between successive terms – the relation has a constant general form. It is this which makes it possible to generalise over the whole infinite sequence to prove convergence or otherwise. But in the case of empirical theories we can have no such proof for we have no rectifying *rule* for produc-

[3] To say more exactly what is meant by this is not easy. Examples, frequently cited, of this kind of relationship are the laws of classical mechanics and relativity theory on the one hand, and the laws of classical (statistical) mechanics and quantum theory on the other. Another slightly different and more recent example is provided by the sequence of accounts of the structure of DNA and of the mechanisms of genetic coding, where striking advances have been made in the detailed precision of our knowledge (as evidence of which genetic engineering is no longer a matter only for science fiction) but it is also evident that there are many details of this mechanism still to be filled in. This kind of relationship has been treated as evidence of convergence by Putnam (1978) and others. Putnam argues that realism – the assumption that the terms of our scientific theories refer to and describe, in some respects correctly, objects existing independently of them – is the only possible explanation for this convergence. Bachelard's unattainable terminus of a sequence of theories is, by contrast, on the order of a thing-in-itself, undescribable and certainly not described by our theories. It does not explain the convergence exactly, but is something presumed in, or inseparable from, the idea that theories form a converging series (one which continues) in the same sort of way that the idea of a limit is inseparable from the idea of a convergent series – there is no independent route to its existence.

ing the sequence of theories with their concepts. The object, reality, as the terminus of this sequence is implicated in the conviction that the sequence can be continued – that present theories, indeed all theories, by their nature can only be approximations and can therefore be successively improved upon.

It is this entering of physical reality as a purely functional category which is nevertheless essential to the whole continuance of the scientific enterprise and the kind of cognitive enterprise which Bachelard seeks to characterise. He sees contemporary science as committed to a non-idealist philosophy (CA pp.13 and 299). It is committed (without proof) to a transcendent reality (the thing-in-itself) and this enters with the idea of the history of science as revealing a sequence of approximations. The extent to which we can think of science as making objective progress and as legitimately taking objective knowledge as its goal is therefore intimately bound up with the extent to which history can be seen on the model of a *continuable* sequence of approximations. 'Approximation is unachieved objectivation, but it is prudent, fertile objectivation, which is truly rational because it is at once conscious of its insufficiency and of its progress' (CA p.300). In addition (as explained on pp.100–4) the notion of approximation, whilst introducing that of a limit, as that to which approximation is made, does not require knowledge of this limit for its application, since approximation is not judged by comparison with that limit. The question of the right to employ the notion of approximation in the context of scientific knowledge therefore reduces to that of whether the history of scientific concepts can be seen not just as a sequence of rectifications but also as a necessarily non-terminating sequence, one which has a future as well as a past. It is here that the realist commitment is made, a commitment to an unattainable limit, something irreducible which will always transcend our thought of it.

Mathematical 'reality' has a subjective origin in that our sense of this 'reality' arises out of the interference of distinct domains both of which are domains of thought, but which nevertheless resist attempts at unification. The cognitive gap, inducing a sense of the independent existence of an object of knowledge (concerning which there are unanswered questions) is a gap within the rational subject between different ways of thinking about things. (The thought of the rational subject becomes an object to itself because it lacks full self-knowledge, has not integrated all aspects of itself, and is conscious of this lack.) Talk of the 'reality' of mathematical objects is thus metaphorical (CA p.186), for they are not items existing independently of the subject. Yet the metaphor is not to be discarded, so long as it is recognised for what it is, for there is a reality underlying mathematical structures and mathematical activity, namely the rational activity of rational subjects. Mathematics has a future as well as a past, a future assured by the dynamic nature of rational, critically reflective thought which transforms itself in the very act of self-characterisation. Its structures are therefore necessarily open, never fully characterised or known beyond the possibility of extension.

Our sense of physical reality, on the other hand, arises out of the interference of thought with experience and of the resistance of the physical world to our will. This is the interference crucial to the co-ordinate conceptions of self and non-self, grounding the positing of a subject–object distinction. The epistemological endeavours of natural science thus involve an attempt to reconstruct in thought something which is of a quite different character, something beyond the subject. The reality of the physical world is intrinsic (genuine) in that it is from the start posited as something of which our knowledge is, and must be, incomplete. Mathematical objects appear first as mere formal constructs complete knowledge of which seems possible (and even attained). These constructs only take on the appearance of reality when brought into relation with other domains. Their reality is in this sense extrinsic to them. This being so, answers to questions about them are not to be sought in the definitions which originally gave rise to them, but in the construction of new, more general definitions. The answers to questions about the physical world, however, can only ultimately be answered by investigating that world itself, experimentally.

3 INDUCTION, EXPERIMENTAL ERROR AND IDENTITY

Part of what is involved in the resistance of the world to our wills is an enforced recognition that subjective identity (indiscernibility) is not objective identity. (Subjectively) identical acts do not produce identical (desired) results. Enumerative induction over everyday concepts frequently fails. In mathematics, by contrast, subjective identity is the only identity; mathematical objects are *exactly* given by their definitions, or within the systems of conventions which give rise to them. Questions arise about the relations between objects given within different systems and these do not have readily given, determinate answers, but if they are answerable, they are answerable determinately, exactly.[4] This is, according to Bachelard's account, because mathematical domains are domains of rational constructs. All that is necessary for knowledge of the identity of a rational construct is the principle of construction, for the whole *raison d'être* of mathematical objects is to be the reification of principles of construction (principles for the construction of representations); they make it possible to theorise about representation by linking methods of construction with the structure of what is constructed. The only kind of knowledge required here is knowledge on the part of the rational subject of the (subjective) identity of his own actions. Here again it should be stressed that this is not an identity established by pure introspection, but is relative to a

[4] It is perhaps worth noting that the point at which identification problems do arise within mathematics is in the discrete representation of the continuum. To get 'enough' real numbers one has to have more than those generated by law-governed sequences of rationals – one needs lawless sequences. But lawless sequences cannot be individuated on the basis of any finite number of terms. It is just this which is played upon in proofs of the independence of the Generalised Continuum Hypothesis and the Axiom of Choice from the other axioms of set theory and which thereby leaves their status indeterminate. It also means that the structures of constructive analysis mirror, in many respects, the epistemological situation of natural science.

system of conventions intersubjectively agreed (i.e. to a rational practice). Thus what counts as doing the same thing each time (e.g. as adding two correctly in generating the sequence 2, 4, 6...) is (as Wittgenstein 1967 stresses) determined by convention, by what people agree is the result of adding 2 (which is not at all the same as saying that it is arbitrary). This is another way of cashing out what is meant by saying that mathematical objects have only an illusory reality – they have no intrinsic identity.

Because certainty is possible concerning the identity of the operations by which a rational subject produces his rational constructs, when a sequence (such as the natural numbers) is generated by repeated application of an operation to the results of its own application, one can be sure both of the indefinite repeatability of the operation and of the nature of the sequence generated. This is because one knows that the conditions under which the operation is applied do not change, since the application is always to constructs whose identity is assured. Mathematical induction, making generalisation over such sequences possible, is therefore grounded in the essential generality of a recursive rule – the generality implicit in the performance of an operation that one knows how to do, so that if one can do it once, one can do it *n* times (CA pp.217–18). The same thing is done each time, so there is an invariant relation between the terms of the sequence, one which makes possible talk of the whole sequence generated by the repeated application of an operation to a fixed starting point; when the same thing is done each time this introduces a common element in the results.

However, in the application of empirical concepts, even when this application is fixed by intersubjectively agreed methods of experimentally constructing, preparing or observing the objects falling under them, identity is not assured. For one cannot be certain of the identity of the conditions in which the methods are applied. An empirically constructive operation operates outside thought – its results are determined not merely by (the concept of) what is done, but also by the (independent) nature of what is operated on and by the background conditions under which the operation is performed. This means that even when a concept has been correctly applied in a number of cases (when the same application criteria have been employed and the language has been used correctly) we have no right, and can know we have no right, to presume an objective identity between the cases. The sameness is only the subjective identity of acts which follow the rule for the application of a concept. But a concept linked to such criteria may be wholly conditioned by our vantage point. If we have no assurance of the identity of the conditions under which the concept is applied, we have no assurance of the objective similarity of the items falling under our concept. Simple enumerative induction, starting from common-sense concepts, is therefore out of the question as a method of achieving any kind of objective knowledge, since it presumes the objective adequacy of our initial concepts – ones which are conditioned by our initial, subjective vantage point.

This, from another angle, reinforces the initial need to define the object of scientific investigation empirically and systematically, by the use of agreed, repeatable experimental procedures. But the problem, even here, is that reliability in the identification of phenomena is related to the precision of our knowledge (CA p.130), to the sensitivity of our empirical classificatory procedures.

To illustrate this point Bachelard gives the following example: A sequence of experiments may yield the following induction. If pure sulphuric acid is in contact with pure baryta in a solution of pure water, there will be a total precipitation of barium. But one knows that experimentally this law is only ever approximated. Minor differences in results can be put down to impurities in the substances. Moreover, in its inductive formulation it is not a precise law. To be made precise, criteria for the purity of sulphuric acid, for baryta, for water and for total precipitation must be given. Bachelard then asks whether, given that all the necessary experimental precautions (with their implicit criteria for purity, etc.) are taken, one could be sure of the result. And he answers 'Yes, but only on condition that the statement of the result is not made more precise' (CA p.131). This is because the criteria necessary for a more precise specification of the antecedent (criteria for the purity of the substances involved) are independent of those necessary for a more precise specification of the consequent (total precipitation). We have no guarantee that when these more precise specifications are given, the original inductive connection will continue to hold. The inductively established connection is a loose one, holding between phenomena identified within a range of accuracy determined by the experimental methods used in the experiments forming the basis of the induction. For phenomena more precisely defined the connection has to be retested with experiments of an appropriate level of sensitivity. Inductive generalisations are thus essentially imprecise and provisional because they are based on identifying indiscernibles, the relation between which, because of the limited accuracy of any instruments, can never be presumed to be one of absolute, quantitative identity. New methods, more sensitive instruments, will detect differences between these indiscernibles.

This is a theme to which Bachelard returns again and again; the theme that there are epistemological discontinuities at the empirical, experimental, inductive level as well as at the level of basic theoretical concepts. (There are 'ruptures of scale' (FES p.212).) These are the inductive discontinuities between different orders of magnitude; the inductive conclusions drawn at one level of experimental precision cannot be expected to carry through to levels of experimental precision of a different order of magnitude. This forms the occasion for another attack on objectual realism, for he says that the realist presumes that the object being described has quantitatively determinate and exact characteristics, and that the problem is merely to get an accurate measurement of them. Any quantitatively exact statement about an object's momentum, for example, is regarded as significant and as being true or false, even though we may not be able to say which on the basis of our measurements. It is this way of thinking which suggests that the results of induction should be stated in a quantitatively precise form and which will thus obliterate the fact that different concepts apply at different levels of experimental precision. The lesson of quantum mechanics, Bachelard thinks, should be to make us sensitive to this, for here we have an example of the way in which concepts and ways of thinking which serve us well on the macro-physical level have to be discarded, or at least radically modified, when it comes to the micro-physical level. Here it is not, it seems, legitimate to assume that a particle has both an exact position and an exact momentum. Hence we are forced to relate our conception of what it is that we are measuring to the techniques by which we attempt to measure it.

The tendency to state results with a precision not warranted by the experimental techniques which form their inductive base is regarded by Bachelard as the mark of a non-scientific mentality which has pretensions to scientific objectivity. He gives numerous examples. One from the eighteenth century is Buffon's conclusion that the earth was detached from the sun 74,832 years ago and that in 93,291 years it will be too cold to support life (FES p.214). Scientifically, the degree of precision in the statement of results needs to be linked to the sensitivity of the methods used. There has to be a theory of experimental error. This is particularly true if we are to make sense of magnitudes which are far from the range of common-sense experience. Rather than thinking of them by analogy with experienced magnitudes, we need to abstract from them in order to extend both the concepts and the measurement techniques. We have to think of magnitudes in relation to *methods* of measurement, not in relation to qualities of the object measured. ('It is necessary to reflect in order to measure, not to measure in order to reflect' (FES p.213).) The object of empirical investigation is thus essentially related to the methods of measurement used in its empirical characterisation. The object can change its nature when the degree of approximation is changed, for its identity criteria are tied to experimental procedures which have a fixed order of precision. Inductive conclusions concerning the object are legitimate at this order of precision but not at others. With empirical redefinition, relative to new and more precise procedures, previous inductive conclusions must be rechecked. It is experimental method which has primacy here, not the object as an independent given.

To say that the methods of mathematics and of natural science are different because mathematical objects are abstract whilst physical objects are concrete is less than helpful in the absence of any account of the distinction between abstract and concrete objects. Yet, to the extent that epistemological methods are conditioned by conceptions of the object of knowledge, and vice versa, it points one in the right direction – towards the difference between the functional category 'reality' in mathematics and in natural science, and to differences in the application of the concept of identity within this category in these two kinds of discipline.

Mathematical entities, as products of mathematical methods, as rational constructs, are such that the method of construction determines the identity of the construct. The same construction always has exactly the same result. Relations between such constructs, where they obtain, are thus exact relations. But physical entities and phenomena, as defined by experimental methods of identification and preparation are only inexactly defined; the same methods do not always have the same results. It is the possibility of assurance of identity and of assurance of the exactness with which concepts are applied which Bachelard sees as grounding mathematical induction and the possibility of generalisation in mathematics. Whereas the contrasting lack of assurance in the empirical sphere is seen as accounting for the lack of reliability of induction here, and for the difficulties inherent in establishing empirical generalisations. In other words, this contrast is seen as the ground of the differing epistemological situations in mathematics and in the natural sciences. Empirical inductive reasoning, to be successful, requires the objectivation of our concepts, and this can be achieved only through their successive rectification. Classifications have to be made empirically more precise; concepts have to be brought into closer accord with experience, but in order to do this experience itself has to be refined by the use of instruments.

There is a form of realist commitment made in the idea that our concepts can be corrected in such a way that empirical inductions become more secure by rendering our judgements of identity more objective and objectively more precise. It is this which accounts for the difference between the conclusions which Bachelard draws from the problems created for empirical induction by our lack of knowledge of the true natures of things and those drawn by either the Humean empiricist or the objectual realist. Focussing the contrast between mathematics and empirical science on induction and its conditions provides a basis for thinking about the relation between these two kinds of discipline which is quite different from that derived from the contrast between deductive and inductive reasoning. The two ways of making the contrast are not wholly independent, but are the product of looking at the same situation in different ways. The possibility of seeing the situation in the way that Bachelard does is dependent on the conception of objective knowledge outlined in Chapter 2. It already presumes that the object of scientific knowledge is not immediately given in perception and that the objectivity of our concepts, statements and therefore knowledge is not something given, but something to be attained. Lack of objectivity is therefore something which must be taken into account in the epistemology of the natural sciences.

From Bachelard's point of view the epistemological problem is not how knowledge of general laws is possible on the basis of experience of particular instances, but of how our concepts can be objectified so that we have an objective identification of particular instances. To the extent that we have that, generalisation will be justified. Thus he says (CA p.127) that conceptualisation and induction are essentially the same operation. So, since conceptualisation is a necessary condition of experience, induction too is a necessary condition of experience. 'If you do not accord me the guarantee that lead will melt tomorrow, as today, at 327.4°C, you cannot permit me today to construct the concept of lead. Belief in the well-foundedness of inductive reasoning is already necessary to belief in the identity of objects' (CA pp.127–8). The problems lie in the construction of concepts and in the empirical identification of the physical phenomena falling under them. In a theoretical framework where each substance has a characteristic melting point (at standard atmospheric pressure), then if one sample identified as lead melts at 327.4°C, another sample, if it is indeed a sample of the same substance, must melt at the same temperature, for this is built in to the conception of substance identity. This does not, however, mean that a theoretical framework within which substance concepts are defined will never itself need revision. If melting point were found, empirically, not to go along with other criteria of substance identity, then conceptual/theoretical revision would be necessary. But if we are to operate with conceptions of what we are investigating at all, we must also employ some theoretical framework, however vague, which fixes the kind of thing we think we are studying and which leads us to attribute to it an identity which is not exhausted by our present empirical definition of it. The idea that objectivity is not given, but has to be attained, and the idea that there can be no disinfected, metaphysics-free science, are thus intimately connected. Science cannot do without metaphysics any more than it can do without induction, for the metaphysical framework is that which determines the identity of the objects we believe in.

The ability to make and have made progress in empirical precision can hardly be denied to science. But the whole process of empirical classification, however precise and however well instrumented, will appear arbitrary, conventionally grounded in our concern to manipulate and control our environment, if it is not linked to rationally articulated theoretical conceptions. It is by reference to theoretical conceptions, however sketchy they may initially be, that the kinds of criteria used for classification are justified, and that the kinds of experimental procedures used for investigation are legitimated.

4 FROM BACONIAN TO NON-BACONIAN SCIENCE

It is the entry of this theoretical ingredient and its role in empirical investigation that marks the beginning of the departure from purely inductive methods. Beliefs concerning the nature of phenomena to be further investigated are utilised theoretically to justify and provide a rational basis for the methods used in further investigation; investigation which may in the end lead to revision of the beliefs which gave rise to it. There is in this process a shift from descriptive to normative. Belief about the way things are, based on empirically discovered laws, forms a basis for saying how an investigation, with a given aim, should be conducted, for rules for the discovery of new facts (NES p.140), rules relating to the use and design of experiments and measuring instruments. New facts are made possible by the introduction of a technology and a set of techniques based on past experience. It is in this way, says Bachelard, that deduction enters into the inductive sciences, and it is just this which is not, in his opinion, accounted for by inductive empiricists such as Bacon and Mill.

This entry of deduction into inductive science is closely related to the distinction which Bachelard draws between active and passive conceptualisation. It is when conceptualisation becomes active that even in its empirical, experimental dimension, science ceases to be purely inductive and starts to become applied theoretical science, where theory shapes both the course of experimental investigation and the interpretation of results. Concepts which are theoretical in origin come to have a role in the construction of objects of empirical study. A maintained alternating current, or a transuranian element, are not empirically given phenomena but are the product of theoretically based techniques for organising phenomena. They represent realisations of a rational, conceptual order; an order imposed on the world not just in the manner of a grid through which experience is filtered and formed in the filtering, but in the way in which materials or events are structured according to a preconceived plan.

Bachelard is aware that in placing emphasis on concepts rather than on laws and their statement he appears to be swimming against the tide of opinion which sees judgements as having priority over concepts (CA pp.21–2). However, he explains that he is not in fact divorcing the concept from judgement, for in his discussion of the development of scientific concepts he is concerned not with passive conceptualisation, which is a matter of habit and does not involve judgement, but with active conceptualisation which is the result of judgement.

This distinction relates to the beginning of the epistemological break between science and common sense. The everyday concepts employed in common-sense thought are picked up as a matter of habit, as a result of conditioning (passive conceptualisation). The principles according to which terms are employed are not explicitly formulated. It is from this habitual thought that science must start by becoming aware of some of the principles according to which it functions – by starting to make explicit the assumptions (judgements) which are implicit in it. Scientific investigations cannot begin without intersubjectively agreed and hence explicit empirical criteria for the identification of what is to be investigated. It needs, at the outset, an agreed empirical classificatory scheme. But the selection of such criteria and the setting up of any such scheme is a matter of judgement, judgement based on past experience.[5] Moreover, it amounts to the adoption of hypotheses, since it requires treating empirical criteria as criteria for the identification of something which is located in a non-empirical framework determining what kind of thing it is. (The criteria for identifying gold are criteria for identifying a chemical substance, for example. Their use carries with it the judgement that all samples satisfying these criteria are samples of a single substance, gold. The content of this judgement is, however, de-

[5] For the complexity of the task of even reaching this stage see, for example Crosland 1962 and Fleck 1979 ch. 1.

pendent on the theory of substances, on what it is to be a single substance. This is perhaps more evident in the case of the relation between symptoms and diseases; as the theory of disease changes so does the account of what a given set of symptoms are symptoms of.) Thus Bachelard says: 'if one considers knowledge in the entirety of its endeavour, one will always have to consider the concept as displayed in a synthetic judgement in action' (CA p.22). The point is that a judgement taken in this way, as a basis for action, no longer results in a descriptive statement linking independent concepts, but in a rule or principle constitutive of the concepts involved. For the time being it is not treated as something which could possibly be descriptively false, and hence not as something which could be descriptively (contingently) true either. In this way the revision of concepts and the revision of synthetic judgements will go hand in hand.

For example, take Newton's second law of motion ($f = ma$). Prior to acceptance of the principle of inertia, this could not even have been regarded as possibly true, since within an Aristotelian framework it is obviously (necessarily) false. In this framework it is not change of velocity which needs explanation by reference to a force, but the mere maintenance of a constant velocity (when the motion is enforced, as opposed to natural). The direction of enquiry concerning the motion of projectiles is thus focussed on what keeps them going rather than on what causes them to stop. After the break with Aristotelian mechanics and prior to the working out of the Newtonian system, force, mass and acceleration were notions with an independent currency, ones whose relation was a subject of debate. The initial proposal of Newton's law thus has the appearance of a synthetic, descriptive statement. But within the Newtonian system this law is something which cannot be denied; for it is a principle used in mechanical reasoning. Acceptance and use of not just this law but of the Newtonian framework thus involves a further modification of the concepts, something which goes hand in hand with no longer regarding the law as having a descriptive function, but as having a normative, modal force, via its role in both mechanical theory and its application (cf. CA p.20). With acceptance of this framework experimental enquiry is focussed on changes in velocity.

5 CONCEPTS AND THE DYNAMICS OF CONCEPTUAL CHANGE

But how is it that Bachelard can think of conceptual revision as a rational process, one by means of which cognitive progress is made, when the very notion of conceptual change has given his analytic counterparts so many headaches? The first stage of an answer lies in noticing that his notion 'concept' is not the same as theirs; his concepts are not the meanings of words in a language. The second stage is contained in the discussion of mathematics, for progress in mathematics, as we have seen, also necessarily involves conceptual changes. The dynamics of reason in its mathematical employment may thus serve as a guide to the dynamics of theoretical reasoning in science.

a. *Fregean and Bachelardian concepts*

It was Frege's belief that the only legitimate notion of thought, one not tainted by psychologism, is that of the sense of a sentence. If communication and objective knowledge are to be possible, without miracles, the sense of a sentence, what it means, must be something which can be grasped by all users of the language in which it is expressed. Understanding a sentence cannot, therefore, be a matter of associating ideas (private mental images) with words. If there is to be even the possibility of knowledge which is objective in the sense of being accessible to all, the significance of the sentences in which it is expressed must be something on which all can agree. The shift of semantic emphasis from the meanings of words to those of sentences was thus seen by Frege as an integral and necessary part of the rejection of psychologism.

At the beginning of the *Grundlagen* Frege lists the following three principles as principles which are to be observed in his enquiry:

always to separate sharply the psychological from the logical, the subjective from the objective;
never to ask for the meaning of a word in isolation, but only in the context of a proposition;
never to lose sight of the distinction between concept and object.
(Frege 1959 p.Xe)

He comments that if the second principle is not observed, one will be unable to observe the first, for one will be forced to take mental pictures or mental acts as the meanings of words.

Given the second principle, it also follows that object and concept are not, for Frege, separately explicable notions. An object is just that which can fall under a concept; it is the reference of a name, an expression which when joined with a concept–expression yields a sentence which is objectively true or false, depending on whether the object does or does not fall under the concept. The simplest assertoric sentences (such as 'Aristotle was bald') are seen as formed by the juxtaposition of two logically and grammatically distinct components, a name ('Aristotle') and a concept–expression ('... was bald'). Objects, unlike concepts, are complete and self-subsistent. It is this asymmetry between object and concept which, in spite of Frege's insistence that the meanings of neither concept nor object expressions can be given in isolation from those of the sentences in which they occur, means that when scientific theories are formalised in Fregean notation, and are not treated purely formalistically, there is an almost irresistible tendency to think of truth as determined by reference to a world of independently existing physical objects (forming the domain of quantification).[6] Each object in being self-subsistent is independent of all others in the domain and of any concepts under which it may happen to fall. This independence is a reflection of the logical representation of all simple assertoric sentences as contingent and as mutually independent.

[6] This tendency is fuelled by use of a Tarski-type semantics for formalised languages. Frege's own semantics being, as Dummett 1973b has shown, capable of a more constructivist or nominalist construal on this point, despite the underlying realism of Frege's thought. It is, for example, to Tarski's semantic conception of truth that Popper appeals for the legitimation of his notion of objective truth.

The tendency to think of truth as determined by reference to an independent world of objects and its attendant picture of the physical world as consisting of a domain of objects between which there are no essential differences (no differences in kind) leads either to reductionism, or to insistence on a strong continuity between science and common sense.[7] If it is admitted that objects of immediate experience and those of our most fundamental scientific theories are very different in kind, both cannot be regarded as constituents of the 'real' world. The question then arises as to which is the real world, that of science or of common sense (cf. Ayer 1973). Either answer will lead to an attempt to reduce statements purporting to be about one of these worlds to statements about the other. The alternative, non-reductionist approach is to leave the notion of 'object', as that of 'objective truth', quite unexplicated; it is simply the formal locus of objective truth determination. Predicates are then introduced to characterise an object either, for example, as a macroscopic object of experience, or as a fundamental particle. Here there can be no gulf between science and common sense; language and thought about both function in exactly the same ways and their objects form part of a single objective reality. There can be no explication of what it is to be counted an object just because objects are not constituted by concepts; no sense can be made of alternative conceptual schemes and hence no sense at all of the idea that conceptual schemes may have a constitutive role vis-à-vis objects. (See, for example, Davidson 1973.)

In Bachelard's eyes this is the metaphysics of uncritical realism. In either reductionist or non-reductionist forms it represents the kind of metaphysical position which he attacks when he criticises realist philosophers of science on the one hand and phenomenalists on the other. There is thus a point at which Bachelard is in direct disagreement with many philosophers of science working within the analytic tradition, for he rejects the framework of their thought, the metaphysics

[7] Popper opts for continuity between science and common sense.

underlying their formal semantics.[8]

Now Bachelard was as much concerned as Frege with the objective character of scientific knowledge, and, like Frege, he banishes mental images from truly scientific thought just because of their essential subjectivity. Concepts, not images, are the medium of scientific thought. His distinction between image and concept has, therefore, to be just as sharp as Frege's and just as firmly insisted upon. Images belong in the subjective, dream world of poetry, whereas concepts belong to the rationally discursive thought of science, thought which aims to be objective. His concepts cannot, however, be Fregean concepts, for they are not defined via their role in determining the truth values of the assertoric sentences in which they occur.

The basic model for his view of concepts is provided by axiomatic presentations of geometry and the change in attitude towards them that is required by the development of non-Euclidean geometries. Bachelard takes it that one can no longer think of the axioms of a geometry as expressing statements which are true or false in virtue of the meanings of the terms they contain; rather, they must be seen as jointly providing implicit definitions of the terms 'point', 'line', etc.[9] In a formal presentation, the axioms of a system determine what inferences can be made in it. The axioms determine the deductive powers of the terms they contain. The conceptual content of these terms cannot, therefore, be treated as determinate outside the context of such a system.

[8] That there is indeed a metaphysical position underpinning the Fregean logico-semantic framework as it came to be adopted, i.e. via Russell, can readily be seen if one looks at the dispute between Russell and Bradley. It has not gone unnoticed within logically orientated philosophy of science that contemporary science, and quantum mechanics in particular, poses problems for this logical framework. There is indeed a vast literature on quantum logic. But the problematic of much of this discussion is set by a particular kind of logical framework. The metaphysics of realism may be questioned, but not that of objects, although there are exceptions (Bohm 1980 for example).

Bachelard's criticism of the Popperian position, in so far as it employs a notion of objective truth presumed to be elucidated by Tarskian semantics, would be that far from divorcing science from metaphysics, Popper has foisted on contemporary science an inappropriate metaphysics – objectual realism.

[9] This was just the move which Frege was so reluctant to make. See his debate with Hilbert in Frege 1971.

This means that Bachelard's concepts are no more self-subsistent than Frege's. They cannot occur in isolation, but only as part of a system of concepts, the relations between which determine the structure of thought and are manifest in the way in which reasoning is conducted within a particular discipline at a particular period. Moreover, the life of scientific concepts is essentially a social life. The principles which determine the legitimacy or otherwise of inferences are the principles governing scientific practices and these are social practices. Such principles structure scientific thought, and hence the scientific mind, but scientific thought is not an activity in which a solitary individual can engage without introducing (at least in his imagination) his fellow scientists as interlocutors. Bachelard's scientific concepts have their life within communities of scientists.

Now different geometries can be expressed in the same language by the adoption of different axioms. In each such geometry terms such as 'point' and 'line' are associated with different but related concepts, or different concepts of point and line are employed. It is because the same word may, in this way, come to be used to express different concepts that Bachelard does not treat the discussion of concepts as part of a theory of meaning, or as part of the philosophy of language. In fact he sees it to be necessary to distinguish sharply between words with their meaning and function in everyday language, and concepts. A word expresses a concept only when functioning in the restricted context of a scientific (or mathematical) theory. In natural language its function is more diffuse and its meaning accordingly more complex, more variable and less precise. Even for scientists, a term such as 'mass' may not be used to signify just the theoretical concept determined by the postulates of their mechanical theory, but may carry in addition historical accretions 'encrusted' on the word. In these accretions there lies a residual but unrecognised subjectivity. They represent, whilst unrecognised, an epistemological obstacle, an obstacle to fully objective thought. They have to be recognised and stripped off if thought about mass is to become objective.

Here is yet another level at which the discontinuity between science and common sense manifests itself. Common-sense thought is thought in natural language with its rich and complex network of functions, associations and images, all contributing to the meanings of its words. Scientific thought is contrastingly austere; the richness and variety has to be stripped away from the words in which it is expressed. Theoretical scientific thought has to aim to be purely conceptual thought, thought which is structured by theoretical postulates.

To see concepts as forming systems in this way; seeing them as being determined by their systematic interrelationships, is to take a holistic view of scientific theories. In this respect, as in his rejection of the place of images and analogies in truly objective, scientific thought, Bachelard continues in the formalist tradition of Duhem. Concepts are essentially formal and abstract (cf. PN p.134). Their formal nature is a reflection of their essential role in principles of reasoning, and is the upshot of turning away from intuition and intuitively determined content. But it should also be remembered that (and how) Bachelard rejects purely formalist accounts of axiomatisation (see pp.72–8).

Does Bachelard, then, in focussing on concepts as opposed to assertions, slip back into an objectionable psychologism (as both Bhaskar 1975 and Lecourt 1975 charge)? The Fregean rejection of psychologism was designed to secure the possibility of objective knowledge. To this, as we saw on p.48, there are two components: there is (1) the requirement of intersubjective accessibility, and (2) the requirement that correctness or incorrectness be determined by reference to the object of knowledge. It is in response to the first requirement that both Frege and Bachelard draw the distinction between images, or ideas, and concepts. Indeed, any approach to meaning which could be seen as concurring with the very general and unspecific slogan, frequently associated with Wittgenstein's later philosophy, 'meaning is use', could satisfy this requirement provided the use in question is public or corporate use. Bachelardian concepts, determined by the reasoning practices of a scientific community, would thus seem to satisfy this requirement. Yet

Bachelard himself rejects mere intersubjectivity in the life of concepts as insufficient to secure the possibility of objective knowledge,[10] for it leads to nothing more than a new form of psychologism (p.76). The divergence between Frege and Bachelard thus comes to rest on their views of what would constitute objective knowledge and hence of what will satisfy the requirement that the correctness or otherwise of knowledge claims is to be seen as determined by reference to the object of knowledge.

Frege focusses attention on the contents of assertoric sentences, sentences which express propositions or thoughts, because he presumes that to have objective knowledge is to know truths, to be able to make true assertions. It is also presumed that for these to be objectively true they must be truths about a realm whose nature is wholly independent of the subject and they must be determined as true or false by reference to this realm. The reasons why Bachelard does not follow the Fregean path have thus already been rehearsed in section 7 of Chapter 2 (pp.53–8) in connection with the relation between subject and object in objective knowledge.

To have scientific knowledge, it is not enough to be able to make true assertions. Nor can a theory be classified as scientific simply on the grounds of falsifiability. For Bachelard the classification of theories as scientific or otherwise is not a possible classification. It is perfectly possible to have a non-scientific knowledge of Newtonian mechanics, or of relativity theory. This would be the case when the theory has been learnt and remembered as a body of facts (possibly to be regurgitated on an exam paper) without any thought as to why, or whether, it should be accepted. One who has scientific knowledge is like the mathematician able to work out, invent and scrutinise proofs. He is to be contrasted with one who can give a proof or solve a problem only on the basis of having memorised the steps, and who could only check the correctness of his memory

[10] Cf. Anscombe 1976 for discussion of a similar point in relation to Wittgenstein.

(and hence his proof) against the book from which he learnt it; he could not say why the proof goes as it does.

The distinction between knowledge of a fact (knowing that something is the case) and scientific knowledge, involving understanding of why something should be the case, is marked by a corresponding distinction between notions of correctness. Facts are said to be *véritable* (true or genuine) whereas theories and laws are *véridique* (veracious, credible). Laws are not just summaries of facts (as they would be if their universal quantifiers were interpreted extensionally), for they introduce modalities; they say what is possible and impossible. As such they cannot be true or false in the same way that factually descriptive statements are. Questions of the objective correctness of scientific knowledge have therefore a double aspect (empirical–rational) (MR p.224).

It is because Bachelard insists on the modal force of scientific laws, a force which indicates their role as principles of scientific reasoning, that the quest for better theories and laws can be seen as a quest for better concepts, that scientific progress is seen as requiring the rectification of concepts and conceptual systems rather than statements. For example, mass, space and time are fundamental concepts of Newtonian mechanics. All objects with which the theory deals have mass and are located within space and time. From the point of view of this theory, an object without mass is an impossibility, as is an object which is in two places at once. There are thus statements involving these concepts which are necessarily true, statements such as 'Two objects cannot be in the same place at the same time' and laws such as the conservation of mass. These principles form part of the framework for identifying and individuating material objects – the objects of classical mechanics. They are constitutive principles relative to this domain, and as such cannot be denied if we are still to be talking of material objects.

On this conception of a scientific theory, to change a theory is not merely to reject as false some belief previously accepted as true. It is to change one's whole way of thinking. The conceptual framework has to be reformed. A mark of such a change will be a change in the modal status accorded to a statement. Something previously regarded as impossible may be countenanced as possible, or vice versa. A concept having a constitutive role is 'dialectised' when one becomes conscious of its role and questions it by asking whether one of these necessary principles really holds. 'Is mass really always conserved?' In the process one is already shifting the status of the principle from necessary to contingent, and this cannot be done without altering (deforming) the concepts involved and the conception of the entities which are the objects of the theory. If conservation of mass can be questioned, the objects cannot be just quantities of matter. But this now requires a new conceptual framework within which to discuss the question.

Such questioning is one of the mechanisms of scientific change. It is one which operates on the theoretical plane. It is intrinsically related to the idea that the aim is to achieve increasingly objective knowledge. Progress toward objectivity at the theoretical, conceptual level is a matter of gradually revealing the uncritical assumptions built in to the way in which we think about the world. The break with common sense is (apart from the switch in epistemological values involved in adopting the aims of science) not made all at once, but gradually, nor is it ever completed. To question a presupposition is to attain a vantage point from which it can be questioned. This will be a point affording a more objective view if it encompasses the previous position (i.e. if it is not just an opposing view incorporating only the negation of the principle in question). This will be the case if a more general theory is achieved in which it is possible to talk both of domains in which the principle (and hence the old theory) holds and ones where it does not. The more general theory has not only a wider domain of possible application, but is independent of some of the presuppositions implicit in the framework of the old theory. It thus provides a vantage point from which the old theory can be evaluated and the limits of its legitimate application set.[11]

The requirement for theoretical change to be progressive is

[11] Cf. Einstein and Infeld 1961 pp.151–2.

therefore that it be possible to incorporate within the new theoretical framework a view of the theory rejected or superseded. The new theory has (negatively) to evaluate what preceded it not by contradicting it but by showing where and why it was wrong, by showing its errors, explaining and rectifying them. This will not always mean incorporating the preceding theory as a special case (as happens with classical mechanics in relativity theory). The rejection of phlogiston (and any other theory in which heat is treated as a substance) by kinetic theory does not do this, but nevertheless provides reasons for not regarding heat as a substance. From this vantage point one sees that heat cannot be a substance. It thus provides a basis for a negative evaluation of phlogiston theory.

Theoretical change satisfying this requirement is directional; it has a direction built into it. A change from relativity theory to classical mechanics would not satisfy the requirement, and therefore would not be regarded as progressive. It is also conceptually linked to one aspect of the notion of objectivity: attainment of a neutral view, and so has some claim to be regarded as development in the direction of objective knowledge. Once such a change has occurred, regress to the previous position is impossible. The situation is much like that on which Gombrich comments when he says of modern artistic attempts to recreate primitive images:

> Whatever the nostalgic wish of their makers, the meanings of these forms can never be the same as that of their primitive models.... No sooner is an image presented as art than, by this very act, a new frame of reference is created which it cannot escape.... If ... Picasso would turn from pottery to hobby horses ... he could not make the hobby horse mean to us what it meant to its first creator. That way is barred by the angel with the flaming sword.
>
> (Gombrich 1963 p.11)

A new theory is a new frame of reference. The perspective of a previous reference frame can never be regained. Even if the contemporary position is, in its turn, rejected along with its way of evaluating its past, it will have to be superseded by a theory which can evaluate it and its evaluations of the past. The past may be re-evaluated, but not regained.[12]

Because a change of theory involves both a change of concepts and a change in the objects of thought employing those concepts, it entails a certain conceptual discontinuity, and the objects of scientific theorising will, to the extent that they have different essential properties, be different objects. This seems to assimilate Bachelard's position fairly closely to those of Feyerabend and Kuhn. The holistic view of concepts and theories looks as if it can lead merely to corporate, socialised idealism, even if conceptual development can be rationally justified and has a historical direction. It looks as if it leaves concepts, and hence scientific theories, as objective only in the sense of being publicly expressed and acknowledged. Access to any objective reality, or to any world of self-subsistent objects, seems blocked by allowing concepts a constitutive role in scientific thought.

It was for just this reason that Frege resisted taking such a view of the axioms of geometry and that Russell claimed that implicit definition, via sets of postulates, has all the advantages of theft over honest toil (Russell 1919 p.71). One can in this way define any concepts one wants, but this does not ensure them any application, or give any indication of how they are to be applied. In this sense implicitly defined notions lack objectivity.

However, what has been omitted from this discussion of the theoretical/conceptual dimension of scientific thought is its relation to experiment and observation. Bachelard, like Popper, links the objectivity of science closely with its empirical character, although the connection is made in a different way. Because, on Popper's view, deductive relations can hold only between sentences, of the statements they are used to make, theory and experience can be deductively related only via language. If concepts are Fregean concepts and provide the meanings of predicate expressions of the language being used, and if the content of these concepts is allowed to be implicitly

[12] Cf. Fleck 1979 ch. IV for the corresponding phenomenon at the empirical level.

determined by the postulates of whatever theory is in question, there can be no possibility of a deductive relation between this theory and the world which could at the same time serve the critical function required by Popper. But Bachelard does not think of inferential relations as necessarily being relations between sentences, or statements; inference is not necessarily linguistically expressed. Nor does he think of concepts as giving the meanings of words in a language. Rational thought is manifest not only in verbal reasoning, but also in the practical reasoning which goes into the design of experiments and the construction of instruments and machines.

Because rational thought is not restricted to pure theory, the practice within which a scientific concept should be seen as functioning is not a purely linguistic practice; it is not a practice divorced from experiment. This is one of the things which differentiates science from pure mathematics; genuinely scientific thought has to be subject to constraint by the resistance of the material world to its projects (MR p.58). It is via their systematic role in theoretical and experimental practices that scientific concepts are implicitly defined. If coherent concepts emerge from such a practice, they will not be ones whose applicability is in question. The goal of scientific theorising is the development of concepts which can be realised in the material world and it is in the realisation of theoretically defined concepts that the transition to non-Baconian science is made. One cannot realise just any theoretically constructed concept, not all of them are genuine empirical possibilities.

Thus the primary function of a Bachelardian concept, the function by which its content is manifest, is not determination of the truth or falsity of factually descriptive statements, but determination of the correctness or incorrectness of scientific procedures, whether theoretical or experimental. These concepts are therefore, like Frege's, defined via their role in judgement, but not in judgement as to what is or is not actually the case, but in judgement about what is possible or impossible, and thus also about how one could or should proceed in experimental testing, measuring or deductively developing a theory. Such judgements are manifest not merely in what is said, but in what is done, in practices and procedures adopted. This shift in thinking about concepts is a reflection of the shift of attention away from objects, things, and toward methods. ('Science does not correspond to a world to be described. It corresponds to a world to be constructed', ARPC p.65.) Scientific concepts have their life in connection with the methods used within scientific disciplines and with the standards applied in ensuring that the prescribed methods are adhered to. Since these methods have both theoretical and empirical dimensions, so too do scientific concepts. If objective knowledge is to be possible in science, on this account, there thus have to be mechanisms for the revision of practices, of the standards employed, and of the modal statements expressing these, in which they are corrected (rectified) by reference to the object of knowledge. It is this which an account of the dynamics of conceptual change will be required to elucidate.

b *Conceptual dynamics*

The two dynamical factors in mathematical thought were said to be the demand for rigour on the one hand and for unification on the other. The demand for rigour was seen as arising in the face of paradoxes, problems raised by informal mathematical practices. The demand for unification was seen as ensuring that practices would not remain formal and rigourised, but would be extended to new domains, 'interference' with which generates new problems, and so on. (See pp.81–90.) The demand for rigour involves a rejection of reliance on intuition and a critical, analytic scrutiny of the assumptions implicit in the non-formal, problem-generating practices.

Bachelard's picture of the dynamic factors in scientific thought is closely analogous. In science, just as in mathematics, there is an initial break with common-sense thought. There has to be a reversal of epistemological values in which the self-evident, the intuitively obvious, is devalued. This devaluing of intuition is analogous, in its dynamic, epistemological function, to the demand for rigour in mathematics. It involves a similar exercise of critical reflection, prompted by encountering difficulties which seem irresoluble within the framework of existing practices. The basic unifying drive, the desire for objective knowledge of the physical world, is that which demands the unification of experience and theory in a coherent system. This involves two kinds of unification: the coherent unification of empirical and theoretical conceptions of a single object of scientific study, and the incorporation of theoretical conceptions within a system of the widest possible application (i.e. in a world-view), one which unifies a diverse range of experimental fields.

In mathematics the initial lack of unity was between discrete and continuous magnitudes. Unification leading to rational comprehension here required an approximative reconstruction of the continuous in terms of the discrete, culminating in the construction of the real number system as a discrete model of the linear continuum and in an axiomatic analysis of the structure required for a set of points to be counted a Euclidean space. In any science, the initial lack of unity is between theory and experience, between the object of investigation as given by empirical criteria and an imposed, rarely precise, conception of what it is that these criteria pick out (cf. pp.121–6). For objects of scientific knowledge are not determined one-dimensionally in terms of empirical methods alone; they have to be regained as objects defined within a theoretical context.

A theory developed to account for a limited range of phenomena will tend to have its application extended beyond that range, and such attempts at extended application will not only throw up new problems but will also result in a modification of the theoretical concepts previously employed, so that the theoretical perspective on the range of phenomena for which it was originally developed will be altered. (An example would be provided by the theory of wave phenomena, with the extension from water, to sound, to light, to wave mechanics. See Hesse 1966.) This process is very similar to the way in which the concept of a circle was altered under the transition to co-ordinate geometry (see pp.96–100). Traditional geometry had to be rethought, but not discarded; its problems took on new forms.

Here again there are two dynamical factors, interplay between which leads to a continuing cyclical development. So, to the extent that these unifying drives are recognised as operative within scientific thought, as incorporated within its standards of objectivity, one will be justified in seeing the resulting sequences of scientific theories and concepts as forming non-terminating sequences even if there are no rules which generate them. In other words, one would be justified in taking such sequences of conceptual rectifications as approximations to objective knowledge if one could say that or when they were convergent, when there was convergence between empirical and theoretical. As we said on p.129, the extent to which we can think of a science as making progress is intimately bound up with the extent to which its history can be seen on the model of a continuable sequence of approximations. Here we need assurance first of continuability and then of convergence, for only convergent sequences are sequences of successively better approximations to a limit.

We have already seen how one can generate two kinds of directional and continuable sequences of concepts. There is the experimentally based sequence of empirical classifications with increasingly precise application criteria, associated with increased accuracy in measurement making finer experimental discriminations possible. There is also the sequence of ever more abstract and more general theoretical concepts generated by successive moves of dialectical generalisation, in much the same way that successively more abstract mathematical concepts are generated. But it is hard to see how sequences of either of these kinds could be tested for convergence. It is only when one takes into account the fact that the apparently empirical sequence can only be continued by, is only generated with the help of moves in the theoretical sequence, and that inductive discontinuity (pp.134–5) across orders of magnitude means that moves in the empirical sequence will not only prompt but also constrain moves in the theoretical sequence, that one begins to see how the notion of convergence might come to have application here.

However, before we can get even this far, we should acknowledge the fact that we have taken the ideas of conceptual rectification, of the development of concepts, rather too much for granted. For the whole difficulty that has been encountered in attempts to explain the mechanisms of conceptual change is that of knowing how the change is to be characterised. Is it a change in concepts themselves, or is it a matter of the replacement of one concept, or system of concepts, by another? When we say that old theories and their concepts are 'contained in' new theories with new conceptual structures, or are 'regained from a new perspective', what do we mean? In one sense we want to say that it is the same concept, and yet in another it is not, for there has been a change of conceptual framework, and on a holistic view of concepts this seems to entail a completely new set of concepts.

The answer, if one is working with Fregean concepts, is clear. Concepts do not change, cannot be extended, cannot be developed. The only kind of change here is change from one set of concepts to another. But what we have so far said about Bachelardian concepts does not suffice to give a clear answer on their behalf. Their formal role, the context in which they function, has been explained in such a way that we have a criterion of conceptual change, a change in the modal status of a sentence in which the concept is employed. However, the epistemological dimension, the way in which conceptual content is determined and in which, therefore, concepts are individuated, has not been discussed. Here, as mentioned in Chapter 1, the historical dimension is seen to be important.

Bachelard believes that to grasp fully and have a clear understanding of contemporary scientific concepts one must know not only something about the theories of which they form a part, but also something about their history – the route which has led to them. There are two reasons for this. One is that concepts are determined within theoretical frameworks, not in isolation, and the cognitive significance of a theory will not be grasped unless the reasons for holding it are appreciated. For knowledge of theoretical propositions to count as scientific knowledge, it is necessary to have some idea of the problems they were designed to solve and of the way in which they represent, or were seen to represent, an advance over predecessors and competitors. In other words, the cognitive significance of a theory can only be grasped fully when the historical perspective is incorporated into that given by its current employment, and, since the contents of concepts are inseparable from the theory in which they are constituted, this remark holds good for scientific concepts also.

But in addition, words functioning in a natural language, and therefore in the everyday discourse of scientists, frequently carry historical accretions. Their unreflective use in a theoretical context may well introduce presuppositions alien to the theory. For this reason it is important, in Bachelard's eyes, that scientists be aware of the history so that concepts, as determined by their most advanced theorising, are kept in sharp focus, and not muddled by subjective, historical associations. The latter can lead to unwarranted assumptions being made unwittingly, so setting up epistemological obstacles, which can only be removed by bringing the offending assumptions to light. (An example of a case in which this might very well happen is with the use of terms such as 'selfish' by sociobiologists when talking about genes. They may, in their guarded moments, be quite clear that these words are not being used with their customary human, moralistic connotations. But on the other hand, it is hard to keep these separated off, and the separation is one which has not been clearly made by popularisers.)

On this kind of view it is impossible to say that a word in a natural language has a precise or fully determinate meaning. It may play a role in a variety of kinds of discourse, where its various roles may be only loosely related. (In Wittgenstein's terms, it may play a role in a number of different language games between which there is only a family resemblance.)

Moreover, the uses to which any individual can and does put a word are determined by his social and epistemological status. The uses which are for him important determine the epistemological profile of the diverse conceptualisations making up his notion of the meaning of the word (PN p.42). In addition, of course, he may associate images with the word.

It is the fact that words, as habitually used in a complex society, carry both this aura of vagueness and speaker relativity that Bachelard cites as making it possible for there to be conceptual development. The fact is that concepts are not given to us with our language as precise and determinate packages which form the units of a consistent, rationally closed system. This latter is the unachievable ideal of a conceptual framework – of an ideal language for science. It is only *lack* of unity and consistency which provides room for the operation of the dynamic principles dictated by the standards of rationality. This is just as in mathematics, where it was the initial lack of rigour and unity which was crucial to there being the possibility of any significant development at all.

To suggest speaker relativity of meaning has been supposed to introduce psychologistic assumptions which are unacceptable in that they make communication miraculous. But this objection really only applies to a speaker relativity which is based on taking ideas or images (necessarily private items) as providing the meanings of words. As has already been said (p.145), Bachelard distinguishes clearly between images and concepts, devoting separate attention in his works on poetics to the use of language to evoke images. But as far as science is concerned, the only relevant meaning is conceptual, and concepts have a wholly public life in language-related practices. It is true that on the sort of view of language suggested, there may well be imperfect communication between speakers of the same natural language. But this will happen because the social contexts in which they have learnt and customarily use certain vocabulary are different. The central thought behind such a view is that concepts are not determined by any purely linguistic framework, but by the practices in which language plays a part. Mastery of English does not necessarily confer on one the ability to use or understand everybody's use of English; one needs in addition to be able to participate in the practices within which those uses occur. It is for this reason that it is really only scientists, and perhaps those who have learnt to talk to the participants in the practices of a particular scientific discipline, who can claim to have anything like a thorough grasp of the concepts with which it works.

To separate out, explicitly, the various ingredients of the conceptual meaning attached to a given word by an individual, or by a group in relation to the uses to which they customarily put the word, is, in Bachelard's terms, to 'psycho-analyse' the concept. It is a form of conceptual analysis, but it is to be distinguished from logical analysis.

Because the issue of conceptual change has played such a central part in recent debates about the rationality of science, and because conceptual rectification *is* Bachelard's account of the rational development of science – is the means by which science is seen as making progress – it will be as well to spell out his position on the analysis of concepts and their development in some detail. To give enough detail would require discussion of at least as many examples as are touched on by Bachelard in his many volumes, and it could be argued that they would need more detailed discussion than he gives them. But here we will have to make do with a single example.

6 ANALYSIS OF A NOTION OF MASS

In PN Bachelard gives an analysis of his own notion of mass. In giving this example, I do not want to stress the details of the analysis, about which there is considerable room for question, since I am more interested in the general characterisation of concepts and of conceptual development which it suggests. Bachelard analyses his own notion of mass into five component concepts. Discussion of the significance of these divisions will be postponed until the next chapter.

1) *The pre-scientific concept*: This represents the pre-scientific

origin of the notion of mass and is a notion which is part of unreflective, common-sense thought. It is a rough quantitative conception which 'has a high degree of substantial reality'. It is the quantity of stuff, be it food, earth or gold. This notion of quantity is closely linked with the idea that more is more desirable. (Of two identically priced pineapples one will choose the larger and/or heavier, for that way one will get more for one's money.) 'To the hungry child, the larger fruit is the better.... The notion of mass concretises the very desire to eat' (PN p.22). This quantitative notion is, however, vague and contains a latent contradiction because volume and weight on occasion diverge as indicators of quantity of stuff, leading one to say sometimes that size is deceptive. Another ingredient of Bachelard's pre-scientific notion of mass is dynamical and derives from the practical observation that heavier or more massive missiles are more effective. On both static and dynamic counts mass appears as a positively valued characteristic. This pre-scientific concept is thus partially evaluative in character; it is thoroughly embedded in the scheme of human purposes, and thus belongs in an anthropomorphic view of the world. It is neither a unitary nor a strictly quantitative concept, since very small masses tend to be ignored altogether.

2) *The operational concept*: This is an empirical concept strictly related to methods of mass comparison, via weight comparison in particular. It corresponds to the first, empirical stage in defining something as an object of scientific investigation. However, it does not arise immediately in association with any scientific quest. A quantitatively precise and objective determination of weight is afforded by instruments, such as balances. These were, and are, frequently used in the absence of any theoretical justification for that use. For example, a beam

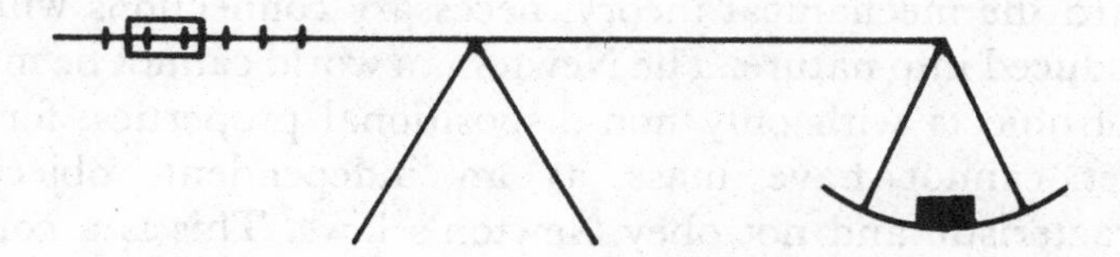

balance with a counterweight sliding along a graduated balance arm was in use long before the theory of the lever was developed.

Such an instrumentally determined concept seems to be both clear and empirically grounded, and is the very model of a scientific concept in the eyes of positivist philosophers of science. But it also reflects a 'realist philosophy' in the sense that it embodies an uncritical assumption that mass/weight is an objective (given) quantity which inheres in objects independently of attempts to measure it. The assumption is that this physical magnitude is a fundamental feature of the world which we merely have to measure with our instruments, but which takes determinate values independently of any methods of measurement. The theoretical justification for the use of measuring instruments is not felt to be urgent because the objectivity of the measure is not put in question.

3) *The Newtonian concept*: In Newtonian, rational mechanics, however, mass is no longer an independently defined empirical notion. It is given only in the context of a body of concepts connected by laws, and in particular as the ratio of force to acceleration in the second law, $f = ma$. Because mass, as so determined, can no longer be treated as a purely empirical concept (since it is no longer a directly accessible sensory quality of objects) it has presented problems to empiricist philosophers (cf. Mach's critique of the concept of mass and of the law of inertia in Mach 1960). However, the formulation of Newtonian mechanics does not require that all three of the notions 'force', 'mass' and 'acceleration' be taken as primitive; two can be treated as primitive and the third as defined (and for discussion of such alternatives see Hertz 1956). A mark of the break between experience and theory which is introduced by rational mechanics is that, if a realist attitude is taken toward the mechanical theory, necessary connections will be introduced into nature. The Newtonian world cannot be made up of objects with only non-dispositional properties, for an object cannot have mass as an independent, objective characteristic and not obey Newton's laws. This is a conse-

quence of making mass into a theoretically constituted concept, one given only in the context of a network of laws; it can make no sense to think of applying this concept in a situation in which the laws do not hold. To this extent an object characterised by its mass (a material body) is a theoretically constituted and identified object; it is one whose identity is not given immediately to sense perception. But there is still a vagueness in the Newtonian concept, for inertial and gravitational mass are amalgamated into a single concept without any theoretical justification.

4) *The relativistic concept*: Within relativity theory, mass, although still a primitive concept, is no longer a simple one. That is to say, although it is not a defined concept, mass is no longer treated as an invariant or essential characteristic of an object. The mass–energy conversion, $e = mc^2$, requires not only that mass no longer be linked to quantity of matter as distinct from energy, but also that mass be functionally related to velocity. A body at rest has mass (rest mass) but not kinetic energy. Relativity theory requires that its resistance to action by an external force (its inertial mass) increase with increase in velocity. But mass is not defined in terms of velocity. This questioning and denial of the absoluteness of mass marks, for Bachelard, the 'opening' of the concept. With the further move to general relativity inertial and gravitational mass become non-accidentally related.

5) *Mass in quantum mechanics*: What is most striking, for Bachelard, about the concept of mass as it appears in the framework of quantum mechanics is that it permits negative mass to be entertained as a possibility. The question of whether a particle can have negative mass is allowed to make sense (or is given a sense) and this in itself marks a fundamental alteration in the concept of mass. 'For the nineteenth century scientist, the concept of a negative mass would have been a monstrous concept. It would have been a mark of a fundamental error in the theory which had produced it' (PN pp.35–6). At this point, he remarks, the philosophy of 'as if', which treats scientific theories as hypothetical fictions, is

shown not to be the philosophy of quantum mechanics, for it could not allow a theory incorporating the possibility of negative mass to be taken seriously. One could not interpret a negative quantity *as if* it were a mass, if the concept of mass is presumed to be defined empirically, independently of the theory. Instead, if one is to follow the attitude of quantum theorists such as Dirac, one must adopt the philosophy of 'why not' and should ask what experiments would, or could, demonstrate the existence of negative mass. One should look for realisations of this entirely new concept, a concept which is without roots in commonsense reality. Bachelard says, in addition, that Dirac's wave mechanics, from the very start, departs from substantial (or objectual) realism, for that which propagates is defined by its manner of propagation, and this in turn is represented in a configuration space, not just in space–time.

7 ANALYSIS OF THE ANALYSIS

Here, then, we have in brief outline Bachelard's analysis of his notion of mass. Its epistemological profile is a rough assessment of the relative contribution made by each of these ingredients. For Bachelard, the Newtonian concept is admitted to be dominant. But, combining this explication of the psychoanalysis of a concept and of an epistemological profile with what Bachelard says elsewhere about conceptual development, more can be said about the relations between the components into which he analyses his notion of mass.

In the first place, it should be remembered that each of the components is a concept, however vague, indeterminate or inconsistent it might be; i.e. each is determined by a range of practices. Each is treated as a concept of mass, and what one wants to know is the justification for this: how can they, from one point of view, be regarded as distinct concepts, and from another be treated as instances of or parts of a single concept 'mass'? This is the problem of conceptual development.

The order of presentation of the ingredients of Bachelard's

notion of mass is historical, so that the quantum mechanical and relativistic concepts of mass are seen as having been arrived at only by a sequence of conceptual rectifications. It is admittedly a sequence constructed with the benefit of hindsight – as a historical sequence it involves an interpretation of history; it is the history of (i.e. which led to) contemporary concepts. The conception of mass as quantity of matter (Newton's definition) is a schematic concept (much like that of a chemical element discussed on pp.44–8) whose exact content depends on the theory of matter (the basic physical theory) adopted. The historical unity in the sequence of concepts is therefore that of location in a structured epistemological field; it is a functional unity.

But Bachelard's point is that this is not just a historical sequence. These epistemological precursors to contemporary scientific concepts still play a role in most people's thought (but not as located in physical theory), for the practices which gave them life still exist. They thus also represent stages in the epistemological development of the individual; the earlier stages begin in some sense conditions of the possibility of development to the later stages. It is important that the scientist be aware of the historical accretions of the word 'mass', not only so that he can be sure to discard them when thinking within the framework of his most advanced theories, but also because these are important for a full grasp of those most advanced concepts. Physical concepts are not fully given merely by their implicit definition in an axiomatisation of contemporary theory, any more than the concept of natural number is fully given by its implicit definition of the Peano axioms. The non-formal, computational practices rigourised by formal axiomatisation are essential to a grasp of the concepts axiomatised, both for a grasp of what applications can be made of them, and for recognition of this axiomatisation as an axiomatisation *of arithmetic* (see pp.79–81). Similarly, recognition of a new concept of mass as a rectification, a rigourising clarification, of a prior, less precisely defined concept requires acquaintance with the more intuitive (subjective) conceptions of mass and the practices within which they are constituted.

Although I do not find the details of Bachelard's account of the pre-scientific concept of mass very convincing, I nevertheless think that the point that there is still a pre-scientific, value-laden, common-sense ingredient in our conception of mass is worth making. It can be brought out by considering Wittgenstein's example of people who buy wood on the basis of the area that it covers; i.e. they pay in proportion to the area covered. One wants to protest that this is not the way to measure quantities of stuff, such as wood, and that it is quite irrational for these people to proceed as they do. 'Why, if I arrange this tall pile into a shorter one that covers a greater area I will get more money for the same amount of wood.' But Wittgenstein imagines that they might respond by saying that if there is a greater area covered there must be more wood, for that is what it means to say that there is more; that is our measure. The deep-rooted feeling that no rational people would adopt such measurement conventions reflects our vague concept of the quantity of a substance. Quantity here is related to value; monetary value is related to other values, in many cases utilitarian – how big a house, boat, table could be built with it? These values presume certain invariances – that the amount of wood does not change with spatial reorganisation. No rational being would (it is felt) accept a measure of wood based on areas covered, irrespective of depth to which it is covered, for it could not be in his interests to do so – he would often have to pay more for what is to him of less value in terms of what he can achieve with, or what benefit he can derive from, his purchase. To make their system appear rational one would have to incorporate into it some account of their society and the basis on which, or the reasons for which, it values wood at all (maybe it has no utilitarian value). The point is that there does seem to be a rough, quantitative, but value-laden conception of the amount of stuff, which still has common-sense currency and which is revealed in intuitive judgements about what are or are not rational ways of assessing such amounts.

Even within such an evaluative framework there is a

pressure to standardise measures and an accompanying tendency to treat quantities as absolute, inherent characteristics of the objects measured. This provides an ontological basis for thinking standardisation to be possible, for thinking that different people and different instruments should be able to agree in their quantitative assessments. Agreed methods of measurement are important in the attempt to ensure fair trading. The introduction of instrumental methods of measurement, ensuring reasonable precision and agreement in quantitative assessments, also forces disambiguation of the vague concept. Yet it is important that whatever measures are introduced should be agreed to be measures of the quantity of a substance (even if different substances have to be measured in different ways). There cannot be complete discontinuity between intuitive and more precise, more objective measurement-based judgements, for purely practical measurement practices and their standards get their point from the value placed on the quantity to be measured and are judged adequate or inadequate in terms of that value.

One is reminded here of the story of how Archimedes came to discover/invent a way of measuring the volume of an irregular solid. The problem, so the story goes, was how to tell whether the king was being cheated by his goldsmiths – how to tell whether they were making his crown using just the gold he had given them for the purpose, or whether they were substituting an equal weight of some base metal for part of it. If the volume of the crown could be determined, one could say whether it was made of pure gold by comparing its weight with that of an equal volume of a sample known to be pure gold. This rests on the empirical finding that equal volumes of different substances have characteristically different weights. The solution, as is well known, was to measure the volume of liquid displaced when the crown was immersed in it. This is an extension of the method for measuring the volume of an irregular container by filling it with liquid and then transferring the liquid to a regular, calibrated or standard container. That Archimedes' method be accepted as a solution to the problem requires that the method suggested be accepted as a way of measuring the volume of an irregular solid, i.e. as measuring the same quantity as other already accepted procedures for measuring the volumes of regular solids. This has to be argued by reference to those procedures. It also requires accepting that knowledge of the volume will lead to an assessment of the amount of gold used in making the crown. This relies on the empirically grounded belief that substances such as gold each have their own specific gravity – the ratio of weight to volume is (roughly) constant for different samples of the same substance. It is this latter concept which, in practical matters, allows not only for the disambiguation of quantity of matter into weight and volume, so resolving the conflicts which arise between them, but also allows for their co-ordination by showing how they can be related, thus distinguishing, but not discarding, ingredients of the previous conception.

A wholly instrumental, empirical concept 'quantity of matter' would have to be treated non-realistically; each instrument or measurement practice would introduce its own concept. But this is not the way of everyday practical, non-theoretical measurement practices. To understand the realist element in the empirical, instrumental concepts it is necessary to recognise their relation to the naively realist, value-laden conception. That the measures are conceived as measures *of something* and that there should be different ways of measuring the same thing is not underwritten by any purely operational concept; the operational concepts are overlaid on (or added to) a preceding vague, but realist, conception to yield more precise realist concepts which have a large empirical, operational component.

Practically, there is a demand to cross-calibrate instruments used for measuring weight by reference to standard weights and standard procedures. For practical matters, agreement in results is what is important. An empirical regularity, such as Hooke's law, relating the extension of a spring to the weight hung on it, can be exploited in the construction of spring balances. The presumption underlying acceptance of a variety of instrumental measures, without other theoretical grounds, is that there is an objective characteristic there to be measured – that the results are not the product of the instruments used, so that on any given occasion one assumes that the result obtained would have been the same if a different method had been used. To the extent that this assumption is neither made explicit nor questioned, and to the extent that empirical regularities are noted, instruments designed on their basis but without asking for explanation of these regularities or for justification of the methods used, Bachelard would not count such practices as genuinely scientific. Even if they are empirically quantitative, they are still pragmatically motivated. Whatever progress there is in the development of methods of measurement, the measures, whilst objective and increasingly precise, are still linked to subjective concerns and to classificatory schemes based on these. Progress can be argued for by reference to existing procedures and justified by utilitarian standards, as the history of measuring instruments shows. But whilst such empirical disambiguation and precision is a necessary groundwork for any scientific enquiry, it does not in itself constitute science, for there is no theoretical dimension – objective knowledge is not the goal.

If one were concerned with the history of theories, rather than with the concept of mass, one would have to note that the theoretical dimension was present long before Newton. The Aristotelian framework provided a theoretical resolution of the volume-weight ambiguity which both does justice to the idea that these are both measures of quantity of matter and explains their relation by explaining why each substance should have a specific gravity. Volume measures quantity of matter (since there is no vacuum), whereas weight measures only the quantity of heavy (or ponderable) matter. Since the character of each substance is determined by the proportions in which the four elements are combined in it, it is natural that they should each have a specific gravity determined by the ratio of heavy to light matter (earth and water to fire and air) that they contain. This theoretical resolution does not survive the demise of the Aristotelian framework so that the relation of the two quantities, volume and weight, to the theoretical notion, quantity of matter, became once again a problem for the natural philosophers of the sixteenth and seventeenth centuries (Descartes, Boyle and Leibniz, for example).

Why then is the Aristotelian concept omitted from Bachelard's account? Because it forms no part of *his* notion of mass. But then how does this square with his insistence that awareness of the history leading to present concepts is essential to a full understanding of those concepts? The answer is that the history of theories does not necessarily yield a convergent sequence of concepts. To explain: on pp.150–1 it was said that the requirement for a theoretical change to be progressive is that it has to be possible to incorporate within the new theoretical framework a view of the theory which is rejected or superseded. The new theory has to be able to show where and why the old theory was either wrong or limited in its applicability (show why it should have been superseded). It was also pointed out that whilst this yields a directional sequence of theories, it does not always mean incorporating preceding theory as a special case of the new theory. Where this does not happen, as with phlogiston theory, or with the Aristotelian theory of elements, the resulting sequence of *theoretical* concepts will not be convergent. Older theoretical concepts may be entirely discarded and new theoretical concepts introduced. A discarded theoretical concept, such as phlogiston, is not rectified and does not form part of the history, given in the form of a sequence of rectifications, leading to any contemporary concept. It will appear only in the sequence of theories, or conceptual frameworks, whose global rectifications have led to present frameworks. There are thus really two distinct unity questions: one concerning the unity of a progressive sequence of theories which makes it possible to say that they are successive theories dealing with a common subject-matter, and the other concerning the unity of a convergent sequence of concepts that make it possible to say that these are successive, corrected versions of the same concept.

The answer to the first question was already suggested in Chapter 2 in connection with the notion of a chemical element. Successive theories work within a structured epistemological field within which certain concepts play a schematic (functional role). They represent categories within the more general functional notion 'reality'. Functionally, they serve to articulate the conception of the theoretical goal by determining the form of theoretical explanation sought for a given range of phenomena (which is not rigidly fixed, but has certain central cases). The unity then is in the employment of some of the same schematic notions in connection with the same empirical, operational identification of the central objects of knowledge.

The history of the concepts which have filled out either the schematic notion 'chemical element' or that of 'quantity of matter' will not look as if it will yield a convergent sequence of concepts. It is only the unity of goal (explanatory ideal) which makes possible the criticism of one such theory from the standpoint of another. Such conceptual non-convergence on the theoretical, as opposed to empirical, level would be characteristic of the early history of any science, for initially the connection between empirically and theoretically determined conceptions of the object of scientific knowledge is so loose that they are virtually autonomous. Quite radical changes in theory could leave the empirical aspect untouched. Not only are the 'theories' largely schematic, they are also radically underdetermined empirically, being more metaphysical than physical and thus, from Bachelard's point of view, consisting largely of projections of subjective categories onto the world. Here the only immediate progress to be made towards objectivity is in the establishment of an empirically precise classificatory scheme or system of measures.

Again, as noted on p. 138, there is a clear sense in which there can be progress within such empirical concepts – progress marked by increased precision and detail. In the case of a system of measures there may be a convergent sequence of empirical concepts as new and more precise methods are introduced by reference to past measurement practices, so building on preceding concepts rather than discarding them. Measures central to widespread practices over which there is a high degree of intersubjective agreement are unlikely to be discarded; theoretical concepts and improved measures have to be defined and justified against their background.

In the case of a classificatory system, such as that for chemical substances, the matter is somewhat different. Substances such as gold, iron and water, which are reasonably common and generally valued, for whatever reason, will form relatively fixed points in any such scheme, being well established as empirical concepts. But here there is a very marked independence between early theoretical views and the empirical concepts. Gold, once thought to be, like all metals, a 'mixt' substance is now reclassified as an element, whilst water, once an element, is reclassified as a compound, but without substantial alteration of the basic empirical concepts. Even when there is in play the theoretical idea that the objective identity of a substance (its nature or essence) is determined by its chemical composition (the elements it contains and the proportion and manner in which they are combined) rather than by its appearance, taste or reactive properties, it is no easy matter to arrive at a classificatory scheme and corresponding nomenclature which realises this ideal. A reading of Boyle, Newton, Priestley or Lavoissier shows just how difficult was the problem of determining chemical compositions and establishing a list of elements. Initial schemes could only be objective in their empirical aspects. And since they were also guided by theoretical conceptions which, as Boyle (1911) was forced to admit, could not get beyond the metaphysical stage and lacked empirical support, some of the changes in classification resulting from changes in theory were quite radical.

As Bachelard sees it, there is a crucial turning point which will occur in the history of any science. This is the point at which theory, the rational organisation of the discipline, takes over from the empirical, instrumental, operational aspect. Subsequent conceptual development is convergent on this level rather than at the empirical level. The crucial point in physics is with the advent of rational mechanics, and the Newtonian system in particular. Bachelard says that in chemistry it occurred with the establishment of Mendeleev's table of elements (see MR chapter III). The mark of having reached this point is the realisation of theoretical concepts. In the case of chemical elements, the full list of elements is not dictated by naturally occurring substances whose initial identification is empirical. Elements such as Rhenium, Francium and Technitium were discovered, or created, because according to the theory they should have existed. Their empirical discovery/creation marks the realisation of theoretical concepts. This means that from the outset the criteria for their identification were theoretical; the empirical criteria used were derived from the theory.

Even prior to the introduction of Newtonian mechanics, machines could be designed by applying the principles of rational kinematics. Mechanical technology is the non-natural product realising the principles of rational mechanics. And one might say that genetic engineering marks the advance of genetic theory into this 'rational' stage. In other words, transition to this stage is coincident with transition, at the empirical level, from phenomenology to phenomeno-technique (to use Bachelard's terms), or from natural history to technologically experimental science. Here the objects of empirical investigation are increasingly abstract–concrete objects (see pp.123–4) increasingly the product of theory-directed techniques. It is in this sense that Bachelard characterises his view of contemporary chemistry as rational materialism, and of the more advanced contemporary sciences generally as rationalism, where this is crucially to be thought of as *applied* rationalism.

From this point on a theoretical change will not be counted as progressive unless in addition to incorporating a view of past theory its (new) concepts admit of empirical realisation. Moreover, the pattern of possible theoretical development is seen henceforth as corresponding closely to the model of theoretical development in mathematics. To have reached this stage at all, the theory has to incorporate basic laws or principles through which its primitive concepts are implicitly defined by their interrelations. Development takes the form of questioning the limits of applicability of one or more of these principles (since their approximate correctness in certain situations is empirically established). The result of such a move is to a more general theory employing concepts of more general application (but less content) of which the present concepts are a special or a limit case. But since other cases will also be allowed for, the working out of the empirical application of the more general theory will be more complex, more detailed, and the experimental realisation of its concepts (sufficiently sensitive to distinguish these from existing concepts) will require greater experimental precision and/or the realisation of new kinds of experimental conditions. Such moves therefore do not involve discarding theories or their concepts (which have already shown their empirical worth), merely their rectification. For this reason a sequence of theoretical concepts so generated will tend to be convergent.

8 CONVERGENCE

But although we have here been drawing a contrast between convergent and non-convergent sequences of concepts and have linked the idea of a convergent sequence to one generated by successive rectifications of a concept, we still have not really said what constitutes such a sequence. All we have said is that in some sense its later members are derived from its earlier members, and that in some sense the earlier members are important for an understanding of, or are in some way a part of, the later concepts.

Going back to the concept of mass, this would be to say, for example, that one cannot understand the Newtonian concept of mass (quantity of matter) without a grasp of the concept empirically determined by the measurement practices associated with balances of various kinds – the measures now treated as measures of weight, but not always distinguished from measures of quantity of matter. As in the case of weight and volume (or for that matter heat and temperature) one needs not only to distinguish the concepts of weight and mass, but also to see how they are related. The Newtonian concept of mass would not have been acceptable as a theoretical explication of 'quantity of matter' if it had departed wholly from the magnitude empirically determined by the use of balances. The theoretical notion is constrained by this empirically successful concept which was already deeply entrenched in commercial and technological practices (involving both its static and its dynamic aspects). The relation between this empirical notion and the theoretical notion 'quantity of matter' can be revised (the empirical measurement operations can be reinterpreted) but not discarded. Thus Newtonian theory argues that even though mass and weight are distinct magnitudes, weight identity is also an indication of mass identity, under the condition that the measurements are carried out at equal distances from the centre of the earth and not too far away from it, since weight will, under that condition, be directly proportional to (gravitational) mass. The theory thus also places limitations on the relation, predicting for example that the readings on a spring balance will, for a given object, vary with distance from the centre of the earth, and that weight comparison between two masses will not always be possible, as in the conditions of weightlessness prevailing in certain regions of outer space.

But equally importantly, the Newtonian concept of mass is not just constrained by its relation to weight measurements, for this relation is also essential to its determination as an empirical, physical concept. For within the framework of pure Newtonian theory, the concept of mass is only mathematically determined. The principles of the theory give only implicit definitions of the primitive concepts and so merely determine them jointly as forming a structure of a certain kind. In particular, for example, mass is fixed as a scalar quantity, with force and acceleration as vector quantities. The forms of these concepts are mathematically determined, but this does not determine anything as regards their empirical application; it does not fix mass as a physical concept. It is its relation to the empirically determined concept of weight and the whole network of practices in which this concept is involved that the Newtonian concept is given physical content. This content is not the result of identification with any pre-existing empirical concept (as positivist thinking would demand) but is the result of being set in a determinate relation to such a concept. This relation does more than indicate the limitations of the extent to which weight-measuring operations can be interpreted as measures of quantity of matter, it provides the basis for *working out* the way in which the concept of mass, as opposed to weight, should be applied in thinking about situations which are already described using the concept of weight. The contrast between the two notions itself suggests extension of the empirical employment of the notion 'quantity of matter' beyond those of weight. The empirical and theoretical frameworks have to be made to interfere and this interference does not leave either of them unchanged. A body of experimental practice and techniques develops around the concepts determined (formally, mathematically) within Newtonian theory, so constituting the Newtonian concepts in a Newtonian framework, which is importantly not just a theoretical framework.

The transition to the relativistic concept of mass can be seen as arising out of a critique of the Newtonian concept by making explicit some of the presuppositions of this framework (Galilean invariance), having reason to question these (Lorentz invariance of Maxwell's equations, Michelson–Morley experiment...) and working out mathematically the structure of the mechanics that would result if the presupposition were abandoned (and were replaced by Lorentz invariance). The relation between the old and the new theories, as definitions of mathematical structures, can be demonstrated mathematically. But the older, Newtonian theory was not just a mathematical theory, its concepts had become embedded in a network of experimental methods, in a whole practice of applied science. The consequences of the mathematically determined (and determinate) theoretical shift have to be worked through to the empirical level. The relativistic concepts have to be developed as physical concepts, and this development is both constrained and made possible by reference to the Newtonian concepts from which they are to be distinguished.

For example, the mechanical consequence of adopting a concept of mass for which $e = mc^2$ holds, rather than one for which $f = ma$ holds, are worked out by reference to the situations to which $f = ma$ was previously thought to apply. The simplest case would be that of a body with definite mass moving in a straight line under the action of an external force in the direction of its motion. If $f = ma$ holds, the velocity of this body will increase without limit and its rate of increase will be proportional to the force applied. But in the new, relativistic, framework this will be so only for relatively small velocities; the closer the velocity of the body approaches the velocity of light, the greater will be the force required to produce any further increment in velocity. It is impossible to increase the velocity of a body moving with the velocity of light; this is constant and maximal. Projectiles moving with velocities close to that of light will thus provide a test of the new theory; these are found at the sub-atomic level of elementary particles, studied with the help of giant particle accelerators.

Because theory and experiment are already inextricably linked in the Newtonian framework (theory structures experimental practice), subsequent theoretical development is constrained not merely by having to contain an account of past theory, but also by the body of experiment shaped by that theory. In giving an account of the limitations of Newtonian theory, that experience has to be reinterpreted, but its existence cannot be denied. The development, therefore, to be counted as genuine progress has to be shown (by mathematical methods) to be an advance in the pure theoretical (rational) dimension – an advance in the direction of unification and simplification of basic concepts – and also has to be shown to be empirically more precise, detailed and accurate in its predictions. The Newtonian framework has to be shown to be adequate up to a certain level of experimental accuracy and under a limited range of conditions (which must include those normal for terrestrial experiments). It has to be shown where the theories diverge in their empirical consequences and requires experimental realisation (confirmation) of some of these divergencies. This requires design, in the light of the new concepts, of new experimental techniques; the creation of a new body of experience.

It is progress which is *simultaneously* progress in these two respects which is counted as progress towards objective knowledge, or in the objectification of concepts. The requirement of simultaneous development in the two dimensions – experimental and theoretical – means that development can only result from the interference of these domains in which each places constraints on, and causes changes in, the other. This interference of theory and experiment is the source of the sense of objectivity in scientific progress; development on each level is constrained by something external to it. When theory and experience are meshed in such a way that the domain of experience is increasingly a domain of constructed, abstract–concrete objects, this interference can be approximately modelled by the interference of one mathematical domain within another. Approximation becomes a more appropriate metaphor; development on this model will result in successive rectification and convergent sequences of concepts.

Crucial to the viability of this account is the extent to which the rational framework of developed sciences can be taken as being mathematical in character. This is the condition of being able to transfer the account of the rational development of mathematics to the situation in the natural sciences. Closely related to this there is a question about the mechanism of the 'interference of domains' – what exactly does this mean in terms of the concepts involved? It is a question about how terms defined in one domain find application in another. In particular this is a problem in relation to the concepts of quantum mechanics, where the important question is how this works not just between abstract domains (which received some discussion, in the form of an example, on pp.91–109) but between the domains of theory and experience. Both of these issues are in turn related to the postponed question of the significance of the divisions which Bachelard introduces into the epistemological profile of his notion of mass.

5

THE EPISTEMOLOGY OF REVOLUTIONS – BETWEEN REALISM AND INSTRUMENTALISM

According to Bachelard, contemporary scientific thought is non-Cartesian, non-Euclidean and non-Baconian. Its epistemology is non-Cartesian in the sense explored in Chapter 2. Its mathematical framework is non-Euclidean, and the importance of this characteristic in shaping Bachelard's picture of contemporary science becomes apparent in the context of the claim, explored in Chapter 3, that mathematics forms the rational framework for scientific thought. Finally, its experimental methods are non-Baconian in that, as was seen in Chapter 4, they do not conform to the inductivist's account of the way in which experience is related to theory. Here the epistemological view is fundamental, as the discussions of Chapters 2 and 3 show. Non-Cartesian epistemology is presupposed in the suggested account of mathematics and its role in science, and in the shift of attention from induction (including under this head both verificationist and falsificationist approaches) to development by the successive rectification of concepts.

It is to epistemology, in the light of the positions outlined in Chapters 3 and 4, that we must now return. For what Bachelard offers is an account of the epistemology of science via an account of the dynamics of scientific thought. The epistemology offered is that of sciences which realise, in unforeseen ways, the seventeenth-century ideal of a science which is at once mathematical and experimental. It is also the epistemology of science in the twentieth century, a century which has seen major theoretical and technological revolutions. If the scientific developments of this century are to be accepted as having constituted cognitive progress, rather than as discrediting the idea that any such progress is possible, then it has to be recognised that scientific progress can involve alterations both in our conceptions of the nature of reality and in our ways of thinking about and investigating it. It is the prospect of a *rational* epistemology of such alterations that has been put in doubt by Feyerabend, Kuhn and others. Yet it is in such alterations that Bachelard finds the central characteristics of contemporary scientific rationality. *'It is at the point when a concept changes its sense that it has the most sense'* (NES p.56). Scientific thought is thus to be characterised dynamically in terms of its repeated attempts at conceptualisation, in terms of the process of rectifying and extending concepts. It is not when working within a fixed set of concepts that reason is manifest, but in the activities bringing about the alteration of concepts. The sense of a scientific concept is not fully given by its location in a given deductive network of concepts, but depends crucially on its relation to concepts which it replaces (including its own, unrectified stages).

It is in the need for twentieth-century science to recognise major conceptual changes as part of the epistemological process that the requirement for a non-Cartesian epistemology is most urgently felt. Cartesian epistemology is foundational; it rests the possibility of knowledge on the existence of a foundation of indubitable, intuitively recognised truths. Revolutionary changes in scientific thought are precisely changes which involve the questioning and demotion of truths previously taken to be intuitively self-evident and beyond question, whether these are observational or highly abstract. Intuitions must be changed; the rational subject must change the forms of his thought and thus the way in which he 'sees' the world as his intellectual view of the basic structure of reality changes.

If twentieth-century science is to see itself as having made cognitive progress it must therefore internalise this process. Just as the epistemology of seventeenth-century science had to internalise the demotion of the unaided senses and the elimination of sensed qualities from physical reality (a demotion required by its rejection of Aristotelian, qualitative science), so the epistemology of twentieth-century science has to internalise the general demotion of intuitive truths (whether of reason or experience) in a recognition that there is no permanent foundation. The process of questioning the apparently unquestionable has to be part of the cognitive process, of progress towards objectivity. For this reason the conception of objective knowledge in play in contemporary science cannot be one based on perceptual metaphors. Objective knowledge is not immediate, but is objective to the extent that the immediate intuitions (dependent on the psychological and cultural constitution of the subject) have been put aside. The prospect of further evolutionary change is built in not merely as a possibility, but as a rational necessity, for it is built in to the very project of producing a mathematical science of nature. This project is uncompletable in two respects, respects which reflect the two-fold nature of the standards of objectivity inherent in the description of the project. A mathematical science of nature has to satisfy both rational and empirical standards. The exact nature of the project is not fixed in detail but assumes new forms because mathematics, the rational framework, has its own dynamics, one which effects revisions at the theoretical level which are fed through to revisions of experimental methods and which thus contribute to the creation of new realms of experimentally produced phenomena. The possibility of transition to a more general, more abstract, less conditioned and limited framework is ever present. The wholly neutral, non-subjective or absolute universal viewpoint is therefore not a possibility. But, in addition, within any given mathematical framework the relation between a rational, mathematical construct and the aspect of empirical being (nature) it attempts to reconstruct can only ever be one of approximation. Empirical methods always carry with them limits of experimental error. Complete adequation between rational representation and the represented empirical domain is therefore impossible, even within a given rational framework; better approximations must remain a possibility.

1 CAUSALITY AND OBJECTIVITY

The divisions in Bachelard's epistemological profile of his notion of mass correspond to just the kind of discontinuous changes (epistemological ruptures) which involve the devaluing of certain intuitions, a questioning of things taken for granted. Moreover, given the functional role of conceptions of reality in defining the object towards which the quest for objective knowledge is directed (see pp.126–31), changes in epistemological values and changes in world view must go hand in hand. The boundaries within the notion of mass, for example, are therefore not just marks of conceptual modification, they mark points at which the concept *had* to change if it was to continue to play a role in structuring the conception of physical reality, if it was to continue to be part of the new world view. Such changes are, therefore, not just changes in the definition of the term (mass) which modify its descriptive content, but are changes in the form of definition, in the character of the concept of mass and hence in the very nature of mass as a physical characteristic.

As we have seen (p.126), the scientific activity of any period takes place within an epistemological field (a problematic) which is structured by its explanatory ideals, its conception of the goal of objective knowledge. The description of this goal can only be schematic in character. It requires a schematic account both of the nature of reality (a metaphysics) and of what it would be to have objective knowledge of that reality. This gives a framework within which questions can be asked, and which determines what can be said to be unknown. The unknown is posited as unknown content but because this content is specifiable only against the background of a *structured* epistemological field, it is not independent of the forms structuring this field, but is constituted as possible knowledge, possible content, by them. The forms of judgement in which ideally objective knowledge can be expressed are thus internally related to the projected structure of reality, for the categories structuring the conception of reality must be those employed in the expression of objective knowledge of it.

Of the schematic notions which are employed in this way, that of 'cause' is seen as being the most fundamental (NES p.115). Just as the goal of science may be schematically specified as the acquisition of objective knowledge, this may be further elaborated by saying that a requirement of this knowledge, if it is to count as scientific knowledge, is that it serve as the basis for the explanation of phenomena. This is merely to distinguish scientific knowledge from empirical knowledge of particular observed phenomena; it imposes the requirement that scientific knowledge be theoretical, but is an otherwise neutral specification, for it says nothing specific about what constitutes an explanation. But one can further say that scientific explanations are causal explanations;[1] to explain scientifically is to give causes. The scientific theories to which appeal is made in giving such explanations are thus theories concerning the causal structure of physical reality.

It is thus possible to say that scientifically objective knowledge requires the development of theories which reproduce, in their rational structures, the causal structures of reality. And this can again be said without reference to any specific view about the nature of either rational or causal structures. What it does mean is that there will be an intimate connection between cognitive ideals, in which the form of the desired theories and explanations is given, and conceptions of the nature of causal relations, the relations seen as structuring physical reality.

The rejection of Cartesian epistemology includes a rejection of the idea that cognitive ideals can be specified once and for all by reference to permanent and immediately knowable rational

[1] Within an Aristotelian framework in which several kinds of cause are allowed corresponding to different kinds of explanation one might simply have equated giving explanations with giving causes. But with the narrowing of notions of cause in natural science to efficient cause there is room for separation between notions of explanation and cause, and thus for debate concerning whether the reasons given in a particular explanation count as causes. Thus, for example, there is the issue of whether functional explanation is causal, and hence whether it is scientifically respectable.

structures. If the structure of explanation could be fixed once and for all by taking logic to determine the possible forms of judgement and/or the forms of reality (depending on whether the laws of logic are regarded as laws of thought or of truth), then the forms of explanation and of theoretical principles could be determined *a priori*. They would be dictated by logic, the complete science of reason. In this case the scientific project would be seen as how to organise experience under these forms; how to reproduce, within the given rational structure, the structure of causal relations. (An example of the puzzles that this can create would be given by surveying that vast literature attempting to give a logical characterisation of causal and counterfactual conditionals.) The basic logical categories would then be the categories under which experience must be organised.

But it is just this sort of proposal which, Bachelard insists, the developments of twentieth-century science require us to abandon. In this insistence Bachelard thus parts company not only with Descartes, but also with Kant and with most analytic philosophers. For he rejects any approach based on a theory (philosophy) of language and meaning according to which the rational structure of *any* language (and therefore of *all* thought) is given *a priori* by the formal logic built into an account of the way in which language functions. Rational structures are not given and thereby fixed once and for all. There is revision of rational structures, revision of cognitive ideals, and accompanying revision of conceptions of the structure of causal relations. Views about the form of theoretical principles have changed.[2] Such revision is simultaneously a revision of rational structures and of conceptions of rationality (cf. PN p.122); it is revision in the conception of what will constitute objective scientific knowledge, revision creating an epistemological rupture.

Applying this to the divisions in the epistemological profile

[2] Note that this would be true even if we were thinking of form only as logical form, for Fregean logic uses forms which are very different from those of traditional Aristotelian logic.

of Bachelard's notion of mass, we find that these divisions can be correlated with changes in views as to the form of the causal relations into which mass, as a genuine and basic physical characteristic, is required to enter.

a *Cause and substance*

The pre-scientific concept of mass is linked with pre-scientific, subjective conceptions of cause, with animistic world views in which acquaintance with causal agents confers personal power (ARPC p.302 and FES chapter VIII). The empirically precise, objective concept of mass is at home in the metaphysics of substance and attribute. It is an objectively determinable, real characteristic of physical objects (substances). Within this framework, knowledge of causes, the ability to give causal explanations, requires knowledge of the natures of substances (independently existing causal agents) and this knowledge takes the form of an account of essence in which the essential attributes of a substance are given. The rational framework is here provided by Aristotelian syllogistic theory, applied in the manner outlined in the *Posterior Analytics*. The form of the principles from which demonstrations proceed is the logical form 'All A's are B'. Explanation is the demonstration of causes, and this demonstration proceeds deductively, logically. The aim is to match causal structures with logical structures, to fit the definitions of substantival terms (definitions expressed in accounts of essence) to the natures of things. This is a framework of objectual realism in which scientific theorising is essentially qualitative. The causal structures of reality are structures of relations between qualities; the identity of the constituents of reality (substances) is determined by their qualities. The structure of rational thought about reality is the structure of logical relations between terms (subjects and predicates).

Newtonian theory does not fit the Aristotelian mould. Not only is the theory quantitative, but its basic principles are expressed as laws, not as accounts of essence. As Bachelard notes (ARPC p.305) there is a radical change when the continuity of time is mathematised and the anthropomorphic notion of cause is replaced by the scientific notion of a function. Newton's laws are expressed in the form of statements concerning the ratios between physical magnitudes (for example, change of motion is proportional to the motive force impressed, and is made in the direction of the right line in which that force is impressed), which have come algebraically to be expressed in the form of second-order differential equations ($d^2s/dt^2 = f/m$, for example). The reasoning involved in constructing causal explanations within this framework is mathematical. More precisely, the rational framework is provided by Euclidean geometry in the algebraised form made available by the use of Cartesian co-ordinates. The forms of causal, functional relations are quantitative and essentially Euclidean.[3] The aim becomes the matching of causal by numerical structures. Causal relations, the relations between one measurable physical characteristic and another, become all important. What is invariant through change is the form of a relation (a complex ratio between physical magnitudes), not the qualitatively identified subject of change.

The agent–patient conception of cause was threatened with eclipse by the causal law, expressed by the mathematical function, and such an eclipse would mark the decline of the substance–attribute metaphysics. But Newtonian science is ambiguous in that it also treats bits of matter as independent existents; (inertial) mass is treated as an intrinsic, invariant and independently possessed quality of a material object. Yet these same objects are also treated as having dynamic properties which are wholly interactive and which can be specified only in the context of laws of interaction. The unexplained identification of gravitational with inertial mass effects a synthesis of these opposed ways of looking at things, and Newton's

[3] Intuitively this is true in the sense that all the physical magnitudes involved and the relations between them are represented spatially and are treated using the mathematics of Euclidean space. More abstractly this means that the relations in question are invariant under the group of transformations that characterise a Euclidean as opposed to a non-Euclidean space.

vacillation on the question of whether the cause of gravitational attraction remained to be given, together with his reluctance to treat gravity as an essential property of matter, illustrates the corresponding junction of explanatory ideals. Gravitational attraction is presented as a mathematical hypothesis, for it is not a cause in either the Cartesian or the Aristotelian sense. If the demands of either the Cartesian or the Aristotelian conceptions of the world and of what constitutes objective knowledge of it were in force, then the cause of gravity remained to be given. The problem is that gravitational action is not something that could be manifested by the body existing in isolation. It is an attractive force requiring at least two bodies, and moreover it does not depend on the bodies alone but varies with their spatial separation. Attractive and repulsive forces have a spatially structured manifestation. The forces, if considered as powers or potentials of material objects, are not powers in the Aristotelian sense, for they are not powers to achieve some particular effect actualised in achieving it, but are continually exercised and constantly actualised with variable effect.

However, in spite of the differences, as Kant shows, it is possible to incorporate the classical Newtonian world within a substance–attribute metaphysics. It is still a world of objects (substances) which have a temporal permanence; they are still re-identifiable subjects of change, even if they are no longer wholly independent existents but are also, essentially, parts of a law-governed whole, the material universe. This aspect of classical physics leads Bachelard to characterise it as still Aristotelian. It still deals with re-identifiable objects (beings) and their states. It could still be said to operate with a logic founded on the principle of identity 'That which is, is' in that it seeks to explain change in terms of the operation of objects which have a permanence and whose identity (whose being) is fixed in terms of the respects in which they remain unchanged (their essential properties). It is this conception of identity which he says (PN p.116) presented an epistemological obstacle to biology, which needed instead to adopt the principle 'That which is, becomes'. But it is also this latter principle which he sees as being required for contemporary physics, which must thus finally depart from the Aristotelian, substantial, objectual mode of thought inherent in common sense and necessary to everyday life.

b *Cause without substance*

The twentieth-century developments leading to relativity theory and quantum mechanics signal two different kinds of change in rational and causal structures. The move to relativity theory requires a revision of the notions of structure and of the way in which fundamental causal structures are to be characterised. Newtonian mechanics required recognition that sense perception cannot, because of its relativity to the sensing subject, be considered to yield direct knowledge of ingredients of physical reality; reality, as that which is sensed, must be described not in terms of sensed qualities, but in terms which are independent of sensation, if it is to be described objectively. Similarly, relativity theory requires recognition that numerical determinations of spatial or temporal separations are also insufficiently objective to be taken as the basis of a characterisation of physical reality, for any such determinations are dependent on the reference frame of the observer. The objectively obtaining causal laws should not be dependent on the state of the observer, but should be specified in such a way as to hold for any observer no matter what his frame of reference (covariance requirement for the general theory).

Attention is thus turned to the invariance conditions of physical laws, i.e. the transformations of their space-time coordinates under which they remain correct (retain the same form). The more restricted condition which the special theory imposes is that the laws of physics should take the same form in all *inertial* frames. In particular, the laws of mechanics and of electro-magnetism should satisfy the same invariance conditions. To specify the transformations under which a set of relationships is preserved is to specify the kind of structure which they define. Structure is now not characterised numerically, but in a generalised geometric way in terms of groups of transformations. New kinds of objectivity demand are here being made and new, non-Euclidean forms of functional relationship are employed for the expression of causal laws. Einstein remarks that the demand that the general laws of nature be covariant with respect to the Lorentz transformations is a precise mathematical condition which relativity theory imposes on candidates for laws of nature (Einstein 1920 p.43). If we are working within relativity theory this is a standard which has to be met by any putative natural law. The metaphysics of relativity theory is that of the field whose structure is quasi-geometrically characterised. Such fields are causal structures, fields of gravitational and electro-magnetic forces.

Quantum mechanics does violence to the Newtonian picture in a different way. Here the new forms of functional relationship admitted for the expression of causal laws makes a direct impact on the notion of cause itself. The assumption that all change is continuous is abandoned, and the allowed discontinuous changes are not precisely predictable, for the laws governing such changes are not deterministic in character. The laws of the micro-physical domain are irreducibly statistical in character, giving only a probability value for the result of an individual measurement. They are not laws governing the change in time of individual objects but laws governing the change in time of the probability of the occurrence of events of a given kind.

But what metaphysical impact does, or should, this have? What modifications does acceptance of quantum mechanics require in the conception of objective scientific knowledge? These are the questions on which there has been so much discussion but so little agreement. The abandonment of determinism at the micro-physical level has seemed to some, including Einstein, to mark a retreat from the conception of science as aiming at knowledge of physical reality. For causality and determinism have been so closely identified that abandonment of Newtonian determinism has seemed to amount to abandonment of causality, the category which plays the fundamental constitutive role in our conception of physical reality. If this has to be discarded, must not the conception of physical reality and with it the ideal of objective scientific knowledge also be abandoned in favour of an instrumentalist conception of scientific theory aiming only to facilitate the prediction and control of phenomena?

Bachelard's opposition to instrumentalism marks his resistance to this line of argument. Whilst he sees the theoretical and experimental practices of micro-physics as embodying a rejection of objectual or substantial realism, he equally sees them as requiring a refusal of the instrumentalist alternative. For this alternative is itself predicated on a prior realism, realism with respect to the immediately observable (see p.52).

It is in quantum mechanics that recognition of the role of mathematics in providing the rational framework for science is required for it is here that scientific theory becomes actually incompatible with the logical structures of subject and predicate and with the metaphysics of substance (object) and attribute (the metaphysics of objectual realism). Neither wave nor particle interpretations are fully consistent with the mathematical formalism. Moreover, the attempt to interpret the formalism as yielding statements concerning the state of a particle at a given time results in a violation of the classical, truth-functional logic of states (specifically the distributive law, $p \,\&(q \vee r) \vdash (p \,\&\, q) \vee (p \,\&\, r)$, is violated).

Yet the move to quantum mechanics cannot be seen, by Bachelard, simply as the replacement of a deterministic theory by an empirically more successful indeterminist one. If this were all that were involved, it could not, by the standards already laid down (p.173), be counted as constituting cognitive progress, for no rational, cognitive justification is thereby given for such a change. No reason is given for thinking that such a change should be thought to lead to a more objectively correct conception of reality. There must be an accompanying critique of determinism. This Bachelard provides, arguing (along lines similar to those adopted by Bohm 1957) that the requirement that the laws appealed to in the provision of causal explanations should be deterministic is one which is subjectively imposed. The universal, metaphysical determinism associated with Newtonian mechanism can then be seen as a projection of this subjective requirement, one which is without foundation. It is not, Bachelard argues, required by or for the application of the concept of cause (NES pp.114–16), which thus survives the abandonment of Newtonian determinism. The more objective point of view is one from which both deterministic and indeterministic causal laws are viewed as possible, and from which limits on the applicability of deterministic forms can be discussed.

These remarks about causality are related to what was said at the beginning of Chapter 4 about induction. The empirical laws at which we arrive by induction from experience, or from sequences of experiments, are initially deterministic in form; they say what *will* happen in a given situation. But they are also initially qualitative; they are not quantitatively precise and indeed have a degree of imprecision built into them via the concepts employed and the margins of error allowed in their application criteria. As the concepts are made more precise, and the experimental conditions are more stringently specified, the accuracy and reliability of the resulting, corrected generalisation is increased.[4] The phenomena thought to be causally related are to a greater degree experimentally controlled and produced; possible interfering factors are excluded as experimental set-ups approximate to causally closed systems. Thus both in the statement of deterministic laws and in the experimental techniques by reference to which they are tested and corrected, the exclusion clause 'in the absence of interfering causes' is appended. A line is drawn both in thought and in experimental technique between the deterministically treated system under investigation and the random interfering

[4] This view would seem to be endorsed by Kuhn, who says: '*The road from scientific law to scientific measurement can rarely be travelled in the reverse direction.* To discover quantitative regularity one must normally know what regularity one is seeking and one's instruments must be designed accordingly; even then nature may not yield consistent or generalisable results without a struggle' (Kuhn 1977 p.219).

causes. The idea that the process of increasing precision in measurement and increasing experimental control can go on indefinitely, so that in the ideal limit there is no random interference, because all factors are known with quantitative precision, subsumed under laws and thereby controlled, is underwritten by the numerically structured, quantitative determinism of the classical mechanist's world-view. This metaphysical determinism, built into the classical conception of physical reality, is seen by Bachelard as constituting an epistemological obstacle which has to be overcome in the acceptance of quantum mechanics (NES p.112).

The conception of physical reality has, for Bachelard, no positive content. It has a functional role in articulating the goal of science, but it is determined negatively as the limit reached by the elimination and correction of all possible errors. This limit notion is itself determined relative to the method by which the sequence of approximations to it is generated. Quantum mechanics forces a reformulation of this limit notion. The sequence of increasingly precise measurements of any given physical magnitude involves a continual interplay between theory and experiment in the design and construction of measuring instruments. The sequences of increasingly controlled experimental conditions for the investigation and definition of particular phenomena is dependent on the simultaneous increase in precision with which individual magnitudes can be measured and maintained at known levels. Whilst quantum mechanics does not call into question the idea that increasingly precise measurements of an individual physical magnitude can be made, it does question the idea that this can be done for all magnitudes simultaneously. The Heisenberg uncertainty principle ($\Delta p \,.\, \Delta q \geqq h$, where h is Planck's constant, p is the position variable, and q the momentum variable) asserts the non-independence of certain pairs of physical magnitudes. Increased precision in the determination of one member of the pair results in a decrease in precision in the determination of the other. This means that there are limits to the experimental control of variables. The idea that this control could be increased indefinitely to the point where there is no line between the determined and the random is incompatible with the uncertainty principle. The conceptions of the limit reached by the elimination and correction of all *possible* errors (the conception of reality and of what would constitute maximally precise knowledge of it) has therefore to be revised. Complete and exact numerical determination of all physical magnitudes cannot be required, because it is impossible, as is complete determinism. The principle according to which this is an impossibility is nevertheless a causal principle; it asserts the causal non-independence of, for example, position and momentum, although the relation of dependence does not take the form of a numerical function.

As in the case of relativity theory, the alteration of the form allowed for causal laws is such as to mark a move away from modelling causal structures as numerical structures. The quantitative determinism imposed by a mathematical framework requiring all causal laws to be expressed as continuous functions of a real variable (interpreted as the time variable) is replaced by what Bachelard calls a 'topological determinism'. In this case too causal structures are characterised algebraically/topologically even though the 'space' employed is highly abstract. The structures emerge not at the level of individual events, but at the level of classes of such events (and in this sense there is still a sort of qualitative determinism).

The philosophy of this science is, according to Bachelard, that of applied rationalism, where the rational framework is provided by the mathematics of wave mechanics or of infinite dimensional Hilbert space. The sense in which there is objective knowledge here has to be derived from the direct interference of the mathematically formed theory with technologically formed experience. To know in what sense one can still talk of objective knowledge here would be to know exactly how this differs from the instrumentalist account on which the mathematical theory is treated as a purely formal calculating device for predicting and organising observations, for even according to such an instrumentalist view, standards of theoretical acceptability (but not of truth) are set by the way in which the formalism relates to experimental results.

2 APPLYING MATHEMATICS

Bachelard insists that the 'sources of contemporary experimental thought are in the domain of mathematics' (NES p.138). Yet how can experimental thought be in any way mathematically inspired? To attempt to explain his answer to this question it would be best to consider a relatively simple example which he gives in RA, for although the need for this view is urged in the light of quantum mechanics in particular, this view of the nature of scientific thought, if it does correctly reflect contemporary thought, should have a much wider application.

The example concerns the way in which mathematics is involved in developing the concepts of periodic and simple harmonic motion, concepts which are fundamental to thought about wave phenomena. Bachelard stresses here, as elsewhere, that whatever the initial analogical role of images of ripples caused by stones dropped into ponds etc. these images must be discarded if the contemporary scientific wave concept is to be grasped. Understanding this concept and the terms related to it (amplitude, frequency, wave-length) is strictly related to specifically mathematical forms which have both an abstract and concrete life.

a *Simple harmonic motion*

The development of any mathematical physics in which causal laws are expressed as functions requires a mathematical representation of time. Before a function can be interpreted as a physical law, a law of change, one of its variables must be interpreted as a time variable. In co-ordinate geometry time is represented by a spatial dimension and is thus represented, like space, as a homogeneous geometrical continuum. However, in co-ordinate geometry it is also represented as measured, as divided into units and resolved into points each of which is indexed by a number. But before this image of time as a real number continuum can have any physical significance the division into units must be conceived as physically realisable; i.e. time must be measurable. The step from pure mathematical image (the real line) to a representation of physical time is thus mediated by a still abstract but not *purely* mathematical conception of a periodic motion. 'A *periodic motion* is one in which the motion of a body is identically repeated in each of a succession of equal time intervals' (Shortley and Williams 1965 p.247). An empirically realised periodic motion would therefore serve as a basis for time measurement. But this is still an abstract, *a priori* form, an ideal (RA p.188) for it is not an empirically applicable concept in the absence of a determination of equal time intervals, and this itself has been said to require realisation of a periodic motion. The empirical application of the notion therefore rests on the relations between motions which have superficially a regular, recurrent structure. The motion taken to provide a time standard is assumed, until further notice, to realise the ideal, and this is reflected in the mathematical formulation of the laws of its motion. It is represented as being exactly periodic. But the justification for this representation rests on comparison with other motions, not on comparison with absolute time.

Galileo is said to have been watching a lamp swinging in Pisa cathedral one day (in 1581) when he noticed that as the amplitude of its oscillations died down, the periodic time for the motion remained constant (as measured against his own pulse). If such oscillations are to be truly isochronous then there must be an exact relation between the angular displacement at the end of the lamp's swing and the speed acquired at the mid-point; for a greater angle the speed must be greater. In other words, the motion would have to be a simple harmonic motion. 'A *simple harmonic motion* is the vibratory motion of a particle about an equilibrium position where the restoring force (the force tending to restore the particle to the equilibrium position) is directly proportional to the displacement of the particle from its equilibrium position' (Shortley and Williams 1965 p.248). A motion cannot therefore be described as being a simple harmonic motion without giving it a mathematical representation. It is not enough simply to make Galileo's observation. It is necessary to understand the relation between, in this case, angular displacement and angular velocity.

This is what the mathematical theory of the pendulum achieves. The mathematical approach is to consider first an idealised simple pendulum in which a point mass is suspended by a massless, rigid and inextensible string from a fixed point.

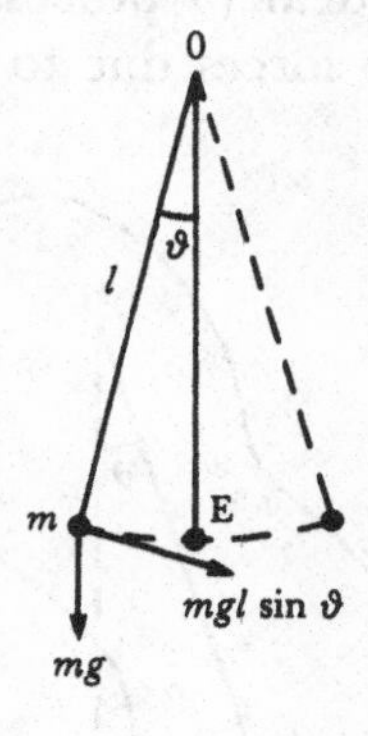

Assuming (as Galileo did, but without any means of precise experimental verification) that gravity exerts a constant force producing a uniform vertical acceleration, we have a torque $mgl\sin\theta$ about the axis through O due to gravity. Since the moment of inertia for this simple pendulum is ml^2, we thus have its period,

$$T = 2\pi\sqrt{\frac{ml^2\theta}{mgl\sin\theta}} \approx 2\pi\sqrt{l/g},$$

for small θ when $\theta/\sin\theta \approx 1$, with θ in radians. The frequency $1/T \approx 1/2\pi\sqrt{l/g}$. So for small displacements, the oscillations of a simple pendulum are approximately isochronous, and the motion is approximately simple harmonic.

The next step towards a physically more realistic situation is to consider the motion of a compound pendulum – a rigid body capable of free motion under gravity about a fixed horizontal axis. This is treated by reducing it to the case of an equivalent simple pendulum by considering the mass, m, of the rigid body to act at its centre of gravity, G. If I is the moment of inertia of the body about the axis of suspension, then for small θ, we get $T = 2\pi\sqrt{I/mgl}$. This formula for the period of a compound pendulum is arrived at by exploiting the possibility of using a mathematical framework to construct an idealised model of the physical situation, and then using the idealisation as an analytical device, seeing actual situations as departures in specifiable respects from the ideal (a process which can go on in stages, e.g. by adding in forces due to air resistance, fiction, etc.).

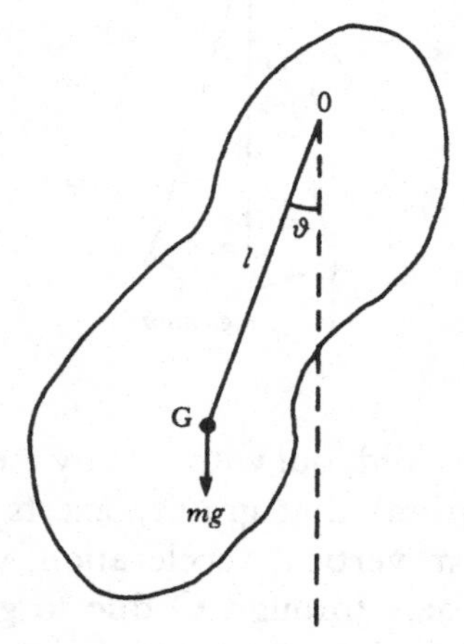

But Bachelard's aim in considering this example is, he says (RA p.188), to familiarise us with the most common wave phenomenon in order to show how the most immediately available example of an oscillation, the swinging of a pendulum, can reveal the structure of relations between fundamental variables. In particular one can show that when the angular displacement of the simple pendulum is given as a function of time, the function is a sine function (a wave function). Similarly a cosine function gives the angular velocity as a function of time. In this way we see the connection between these two magnitudes for the simple pendulum and more generally see the connection between simple harmonic oscillations and wave functions.

Imagine a simple pendulum oscillating (undamped), with a small angle of displacement, above and along the axis of a rotating cylinder in such a way that the bob traces its path on the surface of the cylinder. This trace will be a sinusoidal wave form. If α is the maximum angle of displacement of the pendulum (the angular amplitude) the wave amplitude $A = l\sin\alpha \approx l\alpha$ (since α is small). This thought experiment suggests a physical translation of an angular oscillatory motion into a sine wave and suggests that it should be possible to express this connection more precisely. The period of the pendulum $T = 2\pi\sqrt{l/g}$. This will be the time taken for the bob to go from one crest of the sine wave to the next, in which time the angle determining the value of the sine function will have to have gone through 2π radians. The rate of change of angle with respect to time $\omega = 2\pi/T = \sqrt{g/l}$. At any time t, the angular displacement θ of the pendulum will be given by the function $\theta = \alpha\sin\omega t$ (since $l\sin\theta = A\sin\omega t = l\sin\alpha\sin\omega t$ and for small $\alpha,\theta,\theta \approx \sin\theta$, $\alpha \approx \sin\alpha$). Differentiating, the angular velocity $\theta' = \omega\alpha\cos\omega t$. This allows us to look at the pendulum's oscillations in another way, relating its periodic motion to what is the mathematically simplest and oldest paradigm of periodic motion – uniform circular motion. The ω introduced above is an angular velocity. So consider a point P in uniform circular motion of radius A around a fixed point O with angular velocity ω radians per second. The time t is reckoned from the first time that P passes through the fixed point X.

The angular displacement of P at t is therefore ωt. The y co-ordinate as a function of time is given by $y = A\sin\omega t$, and the x co-ordinate by $x = A\cos\omega t$. As P moves round the circle, M moves back and forth along the diameter YY′. Thus M has an oscillatory motion whose amplitude is A, and whose periodic time $T = 2\pi/\omega$. The motion of M is simply harmonic and the circle on which P moves is termed the auxiliary circle. ωt deter-

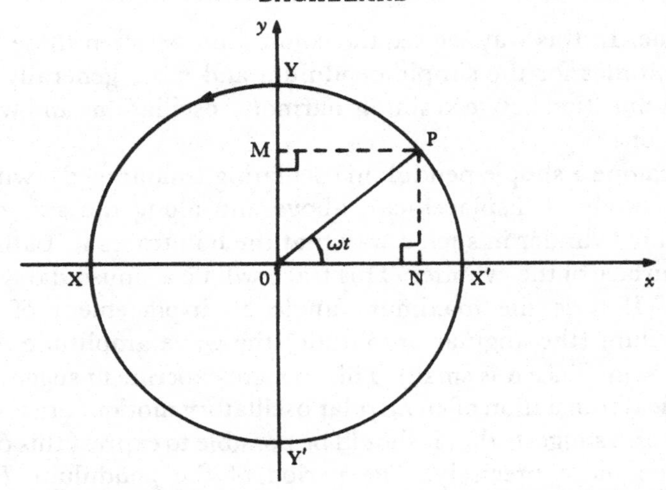

mines the phase of the motion of M at time t and ωt is the phase angle. If $y = A\sin\omega t$ is plotted as a function of time we get the sine wave produced by our pendulum.

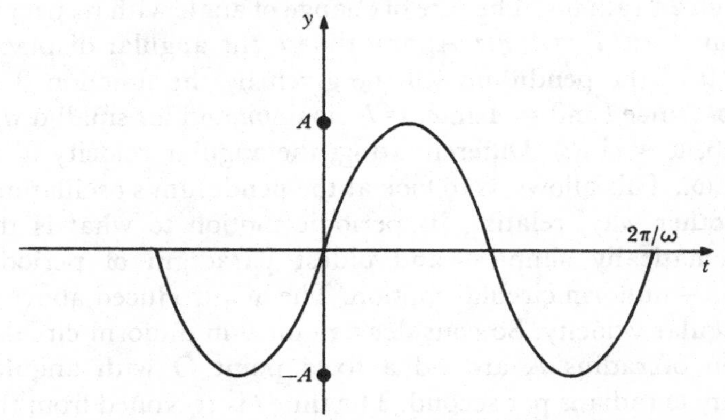

Thus the accelerated and decelerated, continually changing motion of the pendulum is analysed by reference to a uniform angular motion ω, via the trigonometric sine and cosine functions. These functions, therefore, come to play a fundamental role in the study of periodic motions. Vibrations, period, frequency, sine and cosine, Bachelard says, form a complex of notions making possible transactions between mathematics and experience (RA p.190).

Moreover, Fourier's theorem shows that any periodic motion can be considered as the resultant of sinusoidal motions in that it says that any periodic function can be expressed as the sum of sine and cosine terms. This is of fundamental importance from Bachelard's point of view for it makes it possible to take the sinusoidal form as the elementary form of periodic motion, since all periodic phenomena, however empirically arbitrary they may appear, will be analysable in terms of such motions and can therefore receive not merely a mathematical representation, but one which can represent the motion as composite. This is what is essential for the rational framework of an experimental science. It has to be a framework in which causal *synthesis* can be represented, in which the thought experiments of adding in causal factors can be conducted as part of the process of moving from purely theoretical, unrealisable, idealised situations to possible experimental situations. Only a rational framework which makes it possible to articulate concepts constructively can follow science in its work of synthesis. Such a framework will not be blocked by the usual objection that its elementary building blocks are irrational (RA pp.190–4). Simple periodic motions, like uniform circular motions before them, have a very complete mathematical analysis.

b *Construction and the composition of causes*

Now the point of giving this example was to try to get some clearer idea of the way in which there can be interference between mathematical and experimental domains. And this was in order to see if there is a way between instrumentalist and objectual realist views of scientific theories. The first point, then, to note about the discussion of the pendulum and of simple harmonic motion is that, although mathematics is being used quite essentially, it is not being used as a mere formal calculus to predict results about observables. (This is not to say that mathematics is never so used; all that is being claimed is that in its role in physical theorising it is not so used.) The simple pendulum itself is no observable object, but nor is it a mathematical object. It is an object conceived as existing and moving in time; it oscillates. It is an object with mass and subject to a gravitational force.[5] Just as in mathematics, the non-rigorous informal domain which a formalisation is intended to rigourise is essential to an understanding of the formalisation, and in particular for its application, so here the phenomenon of which we are trying to construct a mathematical representation is essential to judgements both of the adequacy of the model and of the way in which it is to be applied. The interference in question here involves an attempt to capture the structure of one domain by constructing a model of it in the other. This can, and must, be a two-way process of trying physically to realise theoretically (mathematically) constructed objects, and trying theoretically to reconstruct physically constructed situations. So, for example, we have the technological problem of how to realise empirically a simple harmonic motion (how to construct an isochronous pendulum). This problem reduces to finding the right curve to put on a pair of jaws which have the effect of shortening the length of the pendulum as the angle of displacement increases. This problem was solved empirically by Huyghens, who also worked out the equation of the curve to be put on the jaws.

For this kind of interference to be possible, it is necessary that there be relations established between mathematical and physical *operations* of composition and transformation (the sort of relation illustrated by the relation between the mathematical projection function and the physical device of marking on a horizontal surface the oscillations of a pendulum, and between the mathematical device of plotting the projection against time

[5] One can advance from the purely kinematic treatment of simple harmonic motion to something more physical, or dynamic, by taking seriously the fact that there is an accelerating mass and that the system is supposed to be a physical system to which the conservation of energy must apply. So for simple harmonic motion we can calculate the energy associated with the oscillating particle by considering its kinetic energy at the equilibrium position O. The velocity of M at O is $\omega A \cos 0 = \omega A$, and since $T = 2\pi/\omega$, $\omega = 2\pi f$, where $f = 1/T$ is frquency, energy $e = \frac{1}{2}(\omega A)^2 m = \frac{1}{2}(2\pi f A)^2 m = 2\pi^2 f^2 A^2 m$. From which one learns that the total energy of a particle of mass m executing simple harmonic motion is proportional to the square of the amplitude and the square of the frequency of the motion.

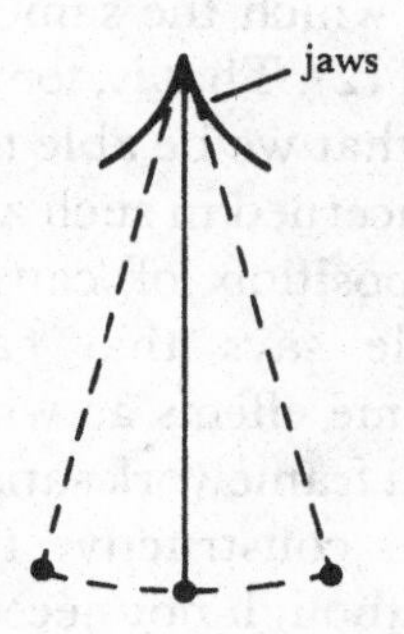

and the physical device of rotating or uniformly moving the surface on which the oscillations are being recorded). In the more general case of applying the theory of wave functions, the operation of functional addition, $(f + g)x = fx + gx$, is given physical significance by the Superposition Principle, which is used in the explanation and reconstruction of interference phenomena. (The principle states that when two or more waves move simultaneously through a given region of space, each wave proceeds independently as if the other were not present, and the resulting 'displacement' at any point in time is the vector sum of the 'displacements' of the individual waves.)

c *Composition of causes and the rational structure of science*

These considerations throw another light on what Bachelard sees as being the crucial turning point in the history of any science – the point at which the rational organisation of the discipline takes over priority from the empirical, instrumental aspect in the determination of concepts. On p.173 it was said that the mark of having reached this point is the ability to realise theoretical concepts. The possession of a rational framework in which the possibilities for causal composition for the range of phenomena in question can be handled is a necessary condition for achieving this stage. As Mill noted: 'a science is either deductive or experimental, according as, in the province it deals with, the effects of causes when conjoined, are or are not the sums of the effects which the same causes produce when separate' (Mill 1971 p.112). That is, for a science to become deductive it is necessary that we be able to analyse the phenomena with which it is concerned in such a way that what he calls the principle of composition of causes holds within that domain. The principle says that causes in combination produce exactly the same effects as when acting singly (Mill 1973 p.376).[6] A rational framework satisfying this demand, i.e. one which incorporates constructive thought, will be essentially mathematical (although not necessarily quantitative) in character, for it is in mathematics that one deals with the principles and forms of constructive thought.[7]

It is not possible, using strictly logical forms, to deduce the result of the co-presence of two or more causal factors from statements of what would happen if each were acting in isolation. This is because the principle 'A ⊢ B entails A,C ⊢ B' is built in to nearly all logical analyses of the relation '⊢' of logical consequence. Deductive inference is frequently contrasted with inductive, causal inference in just this respect. Deductive inference guarantees its conclusion, given the premisses. No further addition of premisses can therefore affect the status of the conclusion. But from the statement of causal antecedents one cannot with the same certainty derive statements of causal consequents because of the ever-present possibility of interfering causes. This is sometimes used (as in Hume) as an

[6] See also Mill 1973 Bk. II ch. IV § 5 and Bk. III ch. VI § 1.

[7] We find Kant making a similar point:

'In order to cognise the possibility of determinate natural things, and hence to cognise them a priori, there is further required that the intuition corresponding to the concept ... be constructed. Now, rational cognition through the construction of concepts is mathematical.... And since in every doctrine of nature only so much science proper is to be found as there is a priori cognition in it, a doctrine of nature will contain only so much science proper as there is applied mathematics in it.

So long, then, as there is for the chemical actions of matters on one another no concept which admits of being constructed, i.e. no law ... can be stated according to which ... their motions together with the consequences of these can be intuited and presented a priori in space (a demand that will hardly ever be fulfilled), chemistry can become nothing more than a systematic art of experimental doctrine, but never a science proper' (Kant 1970 p.7).

Mill similarly notes (Mill 1973 Bk. II ch. IV § 5) that the chemistry of his day does not constitute a deductive science.

argument to show that causal connections can never be deductively established. This is true if deduction is restricted to logical forms (dealing with relations between statements), but not if constructive mathematical deduction is allowed. Causal connections may be deductively established within a theory (may be deductively modelled) without making it possible to predict with certainty the outcome of any given event, because the correctness of the asserted causal connection does not stand or fall with the occurrence of a particular event on a particular occasion, but with the deductive theoretical framework which sets limits on the mechanisms of causal interference and possible modes of causal interaction. Scientific laws have, for Bachelard, it must be remembered, a modal force; they are not interpreted extensionally as universal quantifications over the domain of actual events (see p.149).

3 RATIONAL STRUCTURE, RATIONAL ACTIVITY AND THE FORMS OF EXPERIENCE

However, the required interference between mathematical and experimental domains is not ensured merely by the development of a deductive framework for causal inference. Such interference can only occur as a result of the activity of rational agents. This is why Bachelard insists on the one hand that the phenomena of contemporary science are products of experimental techniques and methods of preparation (whether of the object of study or of the data recorded); they are not naturally occurring phenomena which we just happen to observe, but are the result of our technological intervention, so that what results is structured by the apparatus and methods used. On the other hand, he insists that scientific thought is a structured constructive activity, manifest in the discourse of scientists. Experimental operations structure the phenomena, mathematical operations structure scientific thought.

a *Departing from Kant*

In a way this is Kant's subjective structuring of the world of experience materialised, for the structuring is not just that of the experience of an individual observer, but is physical structuring according to rationally conceived plans, by a group of scientists, where the plans themselves are the product of the sort of mathematically articulated, calculatory thought experiments which are required to mediate between abstract theory and concrete experiment. With the rejection of intuition as a source of mathematical truth, mathematics comes to replace logic in providing the forms of empirical judgements, forms of thought about the empirical world. And with the replacement of perception by elaborately instrumented and designed experiment as the source of phenomena, forms of intuition are replaced by experimental techniques. Yet in spite of this displacement of the Kantian forms, the relation between mathematical and experimental forms is conceived by analogy with the relation in Kant between structures imposed by understanding and by intuition.

The Kantian solution to the problem of how it is possible that mathematical knowledge should be both empirically applicable and rationally demonstrable is to make it knowledge, not of an independently existing reality, but of an aspect of the subject – of the forms of his intuition. It is knowledge possible not *in virtue of* form, but knowledge *of* form (structure). The reflexive potential of the subject for non-intuitive self-knowledge breaks the absoluteness of the form-content dichotomy in thought, making it possible for form to become content to new forms. The forms of intuition need conceptualisation to become objects of knowledge, and because these are forms of intuition, not of concepts, they are forms which structure the content of experience; they are not merely conceptual forms representing items which are independently structured and individuated. The structure of experience is determined by the principles of its synthesis; there is an interplay between the dynamic operation of the principles and the structure which results from their operation. To the extent that these principles can become objects of explicit knowledge there is the possibility of demonstrative knowledge of the structure of experience. What is central here is the idea that experience is the result of activity on the part of the subject, and therefore that these principles are principles of action.

But whereas Kant assumed that the nature of the forms both of intuition and of judgement are fixed, by the nature of the rational subject, and knowable with certainty once and for all, Bachelard, in his rejection of Cartesian epistemology, does not. For even in Kant the idea that the subject should be a possible object of knowledge to itself is problematic. If there is to be the possibility of knowledge of the subject (by himself), if there is to be the possibility of knowledge of his own representational practices and their forms, then, since any such knowledge must modify the subject himself, it must be admitted that there can be no fixed rational representational structure to be known. For the subject to be able to acquire knowledge of his own representational practices and their forms he has already to be able to think beyond them; the acquisition of such reflective knowledge thus modifies the subject (the nature and structure of his thought) in such a way that his reflective knowledge is still incomplete – it is reflective knowledge of the practices he was using in a given domain and of the structures they impose on it, but it is not knowledge of the mode of thought employed in gaining this reflective knowledge. How then should the structure of the situation be characterised?

The most natural suggestion is that it is the structure of a potentially infinite hierarchy of fixed (deductively closed) rational structures each of which could be reflectively known (fully) but only by moving to a wider framework. The fixed and closed character of each of the levels of this hierarchy means that the structure of any level is not altered by the addition of subsequent levels.[8] The element of necessary incompleteness in rational self-knowledge arises merely from the fact that one can never, as it were, catch up with oneself. To think about

[8] Formally speaking the addition of levels is a process of conservative extension.

thought on one level, one has to move to another.[9] (Cf. Chapter 1 pp.25–6) where the very similar argument that criticism must always be external was discussed.) The sequence of levels is open ended whereas the character of thought at each level can be completely known; the history of rational thought is then the history of successively wider, but always closed, rational systems.

But this is not Bachelard's picture, for it is a picture of a succession of static structures which omits entirely the operations generating the sequence; the reflexiveness of rational thought is omitted from its characterisation of rational structure with the consequence that the move to wider structures is not itself represented as a rational process. Yet the whole point of Bachelard's insistence on the openness of contemporary scientific thought is that it has made a second reflection, has recognised the necessary incompleteness both of its rational self-knowledge and of its objective knowledge.

> The scientific spirit is essentially a rectification of knowledge, a widening of the frameworks of knowledge. It judges its past history by condemning it. . . . Scientifically, one thinks of the truth as historical rectification of a longstanding error, one thinks of experience as rectification of an initial common illusion. The whole of the intellectual life of science plays dialectically on this differential of knowledge, at the frontier of the unknown. The very essence of reflection is to understand what one has not understood.
>
> (NES pp.177–8)

Present science does not incorporate past science or the structure of its practices, its thought, unchanged into its wider frameworks. It passes judgement on past science, sees itself as having 'corrected' it. It incorporates past successes and builds upon pre-existing concepts, but not without changing them. Present science thus internalises its relation to past science

[9] This is the structure which emerges with the study of language formalised using classical (Fregean) logic. The self-referential paradoxes require the introduction of hierarchies, if contradictions are to be avoided. The Liar paradox led Tarski to the hierarchy of object language, meta-language, meta-meta-language. . . . Other paradoxes, including Russell's paradox, led to the introduction of the various forms of type hierarchy used in logic and set theory.

and its practices, is conscious of the normative judgements it makes in relation to its past, but it does this without either making it impossible to see how those practices were justified in the light of past science or presupposing that present science or its practices are beyond criticism. It is the internalisation of the possibility of criticism, its incorporation into the rational framework, which is crucial to the possibility of rationally grounding theory change (see pp.64–5). The rational structure of a being capable of repeated reflective thought (of what can not only represent its own representational activities but also reflectively understand the self-modifying character of such a process) must necessarily be incomplete, and, in this sense, open-ended at every stage. It cannot be a fixed or static structure for it incorporates its own dynamics in a recognition of its own incompleteness and of the demand for greater completeness as a rational demand (cf. pp.104–9). There is recognition of the need for future development, a development which will not leave present ways of thinking, including present perceptions of the past, unchanged. Recognition of incompleteness is thus intimately related to recognition of the need for a recurrent history of science (see Chapter 1). The development of science which is taken to constitute an enlargement of the body of scientific knowledge is in no sense a simple cumulative process. The complex rational structure of this process is incorporated into Bachelard's picture by seeing the rational structure of science as being harnessed to that of mathematics.

To claim that mathematics provides the rational frameworks for contemporary science is thus to depart from Kant in two ways: (1) by making a claim about mathematics, that in so far as there is mathematical knowledge it is not intuitive, but rational, discursive, and thus lacks both absoluteness and completeness, and (2) by making a claim about the reasoning and forms of judgement employed in contemporary science, that these are not exhaustively characterisable using logical forms, are not reducible to subject–predicate (or even subject–many-place–predicate) form, but follow mathematical forms and employ mathematical categories, categories which will change as mathematics develops. Scientific concepts will thus not be seen as falling neatly into those of subject and predicate concepts (concepts of substances and concepts of attributes) even when these are extended by relation and quantity. To say that mathematics is not just a language is to say that it introduces not merely new concepts (new vocabulary), but also new forms of judgement (and correlatively, new forms of reasoning) which give rise to new kinds of concepts whose character is determined by their role in these forms of judgement. Thus, for example, concepts such as spin, frequency and amplitude cannot be separated from the mathematical forms in which they participate.

Experience (empirical knowledge) can result only from the application of the categories corresponding to the forms of judgement within structures determined by the forms of intuition; i.e. the categories have to be thought as realised and intrinsically determined within intuition. Kant carries out the process of deducing the resulting concepts and principles in the Schematism and Analogies in *The Critique of Pure Reason.* In the case of Bachelard's treatment of scientific theories, a formally, mathematically expressed theory, determining only formal concepts, cannot express empirical knowledge until its concepts are experimentally realised. The process of deducing how this might be possible is essentially similar to that described on pp.91–104 in connection with the unification of theories of discrete and continuous magnitudes. It is a matter of reproducing the structures of one domain within, i.e. using procedures available within, the other. Because the process is in the two cases essentially similar Bachelard can subsume both under the notion of interference.

b *Making theories empirical*

What is required for the application of mathematical forms is thus not a dictionary or translation manual, but an ability to think possible physical and experimental situations through mathematical forms. When this is done, as in the case of simple harmonic motion, the physical concepts cannot be pulled apart from their mathematical forms. If these concepts are then to have any actual application it is the concepts in their mathematical form that must be related directly to experimental procedures. But the possibility of all this itself rests on the synthetic activity of a rationally reflective subject seeking to unify his empirical and theoretical conceptions of a phenomenon. This element of psychologism, this necessary reference to the subject seeking knowledge, is required both for the possibility of seeing a formalism as being the formalisation of some previously informally adopted set of procedures, and for the possibility of interpreting any formalised conception as a scheme for its empirical realisation. It is the ability to move between action and its conception from procedure to structure and back again, that is crucial to the cognitive enterprise of experimental science.

Concepts as implicitly defined by any kind of formal axiomatic theory, even if it is intended to be a physical theory, so that it includes terms labelled 'mass', 'position', 'time', etc., are not physically significant concepts. They are purely formal concepts only. To yield objective knowledge such concepts have to be capable of being applied and they can be applied only within the domain of possible experience. Beyond these limits they give only the illusion of reality possessed by mathematical constructs.[10] But with abandonment of the idea that there is a fixed set of rational categories which must be phenomenally realised within space and time, the limits of possible experience are not set by reference to the constitution of the perceiving subject; physical theory itself is used to deduce the limitations on the possibilities for obtaining experimental evidence, where these are considered not as mere practical limitations, but as limitations in principle, for physical theories say what is physically possible and impossible.

[10] Cf. Concepts cannot 'be viewed as applicable to things in themselves, independently of all question as to whether and how these may be given to us': *Critique of Pure Reason* A140 (Kant 1929 p.181).

In other words, to ensure that pure theoretical concepts become empirical one needs to give the conditions under which they can have empirical application and so possess empirical significance, become susceptible to empirical testing and to revision in the light of experience. Only this ensures that when we use them we are talking about the physical world and are not departing into the realm of pure metaphysical speculation. This is the view often read as interchangeable with a positivist, verificationist account of the meaning of scientific terms. But it is not the same thing at all. It is not a demand that each concept, if it is to be empirically significant, must be *definable* in observation terms, for there is no admission of any such category as pure observation terms. No terms are allowed to be purely derived from experience. What is required is, rather, that any newly introduced theoretical concept (including a reformalisation or correction of a previously used concept) must be set in a determinate relation to concepts antecedently existing in the body of experimental and theoretical practices. Since the new concept, or network of concepts, may well involve a displacement of antecedent concepts, this determinate relation is required in order that the proposed conceptual/theoretical revision may be thought through to the experimental level, which will not remain unchanged (see pp.176–7).[11]

4 OBJECTIVITY AND THE LIMITS OF THE POSSIBILITY OF EXPERIENCE

But if it is empirically interpreted scientific theory which sets limits on the possibility of experimental knowledge of a given kind of phenomenon, this means that scientific theory itself sets

[11] This process is also described by Kuhn: 'Contrary to a prevalent impression, most new discoveries and theories in the sciences are not merely additions to the existing stockpile of scientific knowledge. To assimilate them the scientist must usually re-arrange the intellectual and manipulative equipment he has previously relied upon, discarding some elements of his prior belief and practice while finding new significances in and new relationships between many others. Because the old must be revalued and reordered when assimilating the new, discovery and invention in the sciences are usually intrinsically revolutionary' (Kuhn 1977 p.226).

the limits of possible experience and in so doing can force changes in conceptions of reality and in standards of objectivity. This is crucial because it makes metaphysical (world-view) revision part of the cognitive process of science. For if, as was said at the beginning of this chapter, the divisions in Bachelard's epistemological profile of his notion of mass reflect changes in conceptions of reality, co-ordinated with changes in epistemological values, then it is hard to see how the sequence of concepts of mass, said to constitute a progressive sequence (a sequence of rectifications), can really be thought to amount to cognitive progress achieved by any sort of rational process. For it involves changes in conceptions of reality and therefore also of what constitutes objective knowledge of it. How can any such change be argued for? Will it not simply reflect a change in value, value having a non-epistemic, non-cognitive origin?

The divisions in Bachelard's epistemological profile of his notion of mass were also (p.186) correlated with changes in views as to the form of causal laws. It is the working through of changes in conceptions of causality which necessarily has consequence for the delimitation of the domain of possible experience. For it is also via causal interaction that experiments yield results; it is by our causal interaction with the physical world that we gain any empirical knowledge of it. The motivation for introducing new forms of causal law (new explanatory paradigms) may derive from empirical prompting, from inability to construct explanations of the required type for a particular phenomenon or range of phenomena (e.g. Kepler's eventual abandonment of the requirement of uniform circular motion for planetary orbits as a result of being unable to make such models fit Tycho's data with sufficient accuracy, data in which Kepler had great confidence, or Planck's reluctant introduction of his constant). But it is not sufficient that predictive success follow such emendation. Some justification of the change which is not just a change of theory but a change of style of theory has to be given if the resulting theory is to be accepted as an explanatory theory. It has to be argued that the restrictions on forms of explanation which the new theory violates are only the result of subjective imposition in the sense that they embody claims which have not been, and could not have been, even within the framework they define, shown to be either true or false by reference to experience.[12] The restrictions cannot then be insisted upon as being objectively required. Nor, of course, can their negations be *required*. But the refusal of a requirement (the denial of a necessary proposition) demands only recognition of the *possibility* of its violation (the possibility of the negation of the proposition).

Here we see that the epistemological effect of such an argument for putting new, more abstract general requirements on the form of causal laws will be to exclude certain kinds of putative information from the sphere of possible empirically objective knowledge. For, since requirements on the form of causal laws are intimately related to conceptions of reality and of what would constitute objective knowledge of it, if a requirement on causal laws is argued to be subjective in origin, then the features linked with that requirement and previously ascribed to reality will (1) themselves shift from being objective qualities of the constituents of physical reality to being relative features whose observation is to be explained by reference to the kind of characteristics now taken to be fundamental, and (2) be such that knowledge of their absolute value or nature is empirically impossible. In other words, there is a change in the epistemological value attached to information concerning them.

For example, the change in invariance conditions for the laws of mechanics means that any measurement of a spatial or temporal separation must be treated as relative (to a frame of reference). But the particular change involved also requires that the velocity of light be constant in all frames of reference and that it be finite and maximal. A consequence of this is that determinations of absolute simultaneity, as well as of absolute spatial or temporal separation, are impossible and cannot therefore even serve as an ideal at which measurement or em-

[12] A strategy skilfully employed by Boyle against both Aristotelian and Paracelsan forms of element theory (see Boyle 1911).

pirical knowledge can aim. From the Newtonian viewpoint one would see this as placing a limitation on our cognitive capacities – absolute simultaneity exists but it can never be empirically determined. From the Einsteinian point of view absolute simultaneity does not exist (is a concept without empirical application); simultaneity has become a feature of events which is relative to the frame of reference from which they are being observed. The inability in principle is not a limitation on what we can know, but is a statement about the space–time structure of the world we are seeking to know. This world being, for science, the world of *possible*, experimentally mediated, experience. The new relativistic world is explored, in the first instance, by conducting thought experiments. These thought experiments have been successful not only in making it possible to apply the special theory, but also in that the response to relativity theory has very largely been to treat its space–time as the space–time of the universe.

But there is a similar situation in quantum mechanics to which the response has been much less uniform.[13] The Heisenberg indeterminacy principle has been, and can be, treated in two ways. On the basis of the classical picture of particles which have at all times determinate positions and velocities, it is interpreted as placing an in-principle restriction on our cognitive capacities. We will never be able to determine both the position and the momentum of a particle with full accuracy at the same time. The alternative is to see the principle as requiring replacement of the classical picture, or at the very least of placing limits on its applicability, for it is not applicable at the level of micro-physics. 'Particles' do not have determinate paths. It is not that there is something that we cannot know; the principle tells us something about the structure of the micro-world. Exactly what is the problem.

Bohr's interpretation, which is again argued for and

[13] Perhaps this is because absolute space–time has always been controversial, whereas the quantum mechanical viewpoint seems to require a questioning of much more fundamental presuppositions, ones which are very deeply rooted in common-sense thought.

explored by the use of thought experiments,[14] suggests that position and momentum cannot be considered as properties of a particle, but are products of an experimental set-up, so that what is measured, or rather the result obtained, cannot be separated, even conceptually, from the experimental situation which makes the measurement possible and which produces the result. Descriptions always have to be of experimental set-ups, not of particles in isolation or separated from the situation in which they are observed. Descriptions are therefore always in this sense relative – a particle can be described only in relation to an experimental set-up, not absolutely (just as it can be described only relative to a frame of space–time reference, not absolutely). The uncertainty principle then emerges as a statement of the incompatibility of certain kinds of experimental set-ups. On this interpretation, then, experimental arrangements produce the phenomena. The observed results and the methods of obtaining them cannot be separated.

Bohr's interpretation is reached by insisting that the categories of classical physics are the only ones in which we can describe the physical world, which is at the same time the world of possible experience. Experience can only be of experimental set-ups and these we describe classically. We cannot further analyse these set-ups and describe what 'goes on' within them. The quantum mechanical formalism, applied to give statistical predictions about the observations made with such arrangements, is not itself physically interpreted, but remains a mere formalism. It is clear that if the quantum mechanical formalism were to be physically interpreted the classical concepts would have to undergo 'rectification'. But exactly what sort of world picture emerges if we think reality through the categories employed by the quantum mechanical formalism?

Bohm 1980 agrees with Bachelard that it is not a world of independent objects; atomistic conceptions have to be discarded. He goes on to suggest what kind of conceptual changes

[14] See, for example, Bohr 1949.

are required. It is in a similar vein that Bachelard talks of the need to move from a linear determinism to a topological or algebraic determinism (ARPC pp.297–8) and from a metaphysics of being to one of becoming (PN p.116), whilst at the same time agreeing with that part of Bohr's interpretation which insists on abandonment of the naive realism of classical physics, which assumed that the properties of matter were 'there' independently of the use of experimental devices. The whole thrust of Bachelard's work is that the Heisenberg principle should not be treated merely as placing a limit on the possibility of knowledge; we should not think that the particle is in a particular state which we can never fully know. But nor should it be thought to make all empirical observation subjective by making it impossible to separate the state of the system under study from the actual state of our knowledge of it. For there is a difference between insisting that we cannot, in general, think of states of a physical system as objectively determinate independently of our experimental techniques, the means by which we observe them, and insisting that we cannot think of any such state as objectively determinate until it has actually been observed and the observation consciously registered. It is to the latter which Bachelard objects as being subjectivist (resting on a subjectivist interpretation of probability), whilst insisting on the former, i.e. insisting that the Heisenberg principle be treated as a principle concerning the structure of physical reality which forces a revision in our conception of it. 'But since in the present volume we have given ourselves the task of arresting subjectivist interpretations, we shall say a word on the eminently objective character of the principle' (ARPC p.289). And later: 'many philosophers seem incapable of taking on board both the *realism* of the Heisenberg principle and its role as a rational principle' (ARPC p.296). It is the link between reality and rational framework which is insisted upon by Bachelard under the name of applied rationalism. For him the reality of micro-physics is thought through an interpreted quantum mechanical formalism.

5 SCEPTICISM OR THE POSSIBILITY OF KNOWLEDGE?

It may be objected that to alter one's conception of reality to fit the latest scientific theory is to make knowledge too easy to obtain. It makes the theory automatically correct. But (1) the move in no way assures the trivial correctness of any particular theory, for the general conception of reality says nothing about the detailed content of laws. It is in the light of changed forms of law that changes are made in views about the nature and structure of reality and how they can be discovered. (2) It can equally be retorted that to hang on to a conception of reality which our best theories tell us is unknowable is absurd, for if that reality *is* unknowable, on what basis do we claim to know what it is like? The situation forces us to recognise that our views as to the nature of reality, in its most general characteristics, constituted in our cognitive ideals (our conception of what would constitute objective knowledge of reality), were not derived from experience but were imposed on it. Any new conception will similarly involve imposed metaphysical elements, but to the extent that the reasons seeming to impose the new features of theories on us are demands of coherence, both rational and empirical, and to the extent that we can from the new vantage point explain the subjective origin of the rejected components of the older view, we may be justified in seeing the move as a move in the direction of objectivity, but not as establishing the framework of the new theory as a fully objective representation of reality; objective knowledge is in fact harder to obtain on this view, and cannot be finally complete.

It should be emphasised that we are not here discussing particular theories, with specific laws, but the situation which arises when a theory seems to impose previously unrecognised limits on the possibility of experimental knowledge while at the same time employing new forms in its expression of causal laws. Such a situation presents problems for science's conception of its goal and prompts a questioning of assumptions which lie at the heart not just of one theory, but of a whole generation of theories. Whilst science can have a goal which it recognises as unattainable but to which it can approximate without limit, it cannot have a goal which its own theories say is in principle unattainable because different in kind from anything which can be attained by the methods theoretically available. Such a limit cannot even be approximated. The world of empirical science can only be the world of *possible* experience, where this field is itself theoretically and ideally delimited and restricted by the nature (as it is given in theories) of the objects experienced. That is, the world of possible experience is not defined relative to sense perception and its limitations, but by theories of the world themselves. If it is correct to say that nothing can travel faster than light, then it is impossible to aim to establish absolute simultaneity relations between events; the ideal limit of accuracy in measurement of even relations is set by the velocity of light signals. Actual apparatus may approximate to this ideal, but not to the other. Ideally accurate knowledge *is* objective knowledge of reality.

Can we not continue, nevertheless, to employ the concept of absolute simultaneity? In two senses we can, for we can discuss classical theory and elaborate on the reality co-ordinate with it, but such a discussion must always have the status of a mathematical hypothesis[15] – it remains a hypothetical discussion. We can also use the concept in everyday discourse where its use is sufficiently imprecise for it to make no difference whether classical or relativistic theories are correct. It is used here in much the way that the concept of a triangle is used; we recognise that there is necessarily a degree of approximation (error) in the application of the concept, one which, for the purposes in hand, does not matter. Where we cannot use it is in connection with scientific experimental investigations where the level of accuracy is such that the difference between the two frameworks becomes empirically significant. The classical concepts cannot continue to have application in our thinking about experimental investigation at that level of accuracy.

[15] In the sense in which Galileo was asked to retract his view that the earth really moves, and retain it merely as a mathematical hypothesis.

Given that space–time separation is not reference frame invariant, there is no reference frame relative to which absolute simultaneity can be defined, let alone determined. The notion could have application only for a being who could identify and describe times and places without needing a frame of reference. Such a being is analogous to Kant's pure intellectual being who does not need to experience things in space and time. If such a being is possible, his mode of cognition is nevertheless one of which we can make no sense; our concepts have to be applied to things in space and time. To yield objective knowledge concepts have to be capable of being applied and they can be applied only within the domain of possible experience. Beyond those limits they give only the illusion of knowledge, knowledge of an illusory reality (the reality possessed by mathematical constructs) (cf. CA pp.114 and 179), which is in fact nothing more than knowledge of our own constructions.

A consequence of not modifying the conception of reality, and hence of what will count as objective knowledge of it, in a situation where scientific theory itself makes such knowledge impossible to acquire, is that either science must give up its goal of objective knowledge and retreat into some form of instrumentalism, or it must hide from itself the consequences of its own theories. If the image of science as making progress towards objective knowledge is retained, but with an inappropriate conception of what that knowledge consists in, then the image will be ideological, requiring a false consciousness to sustain it. It is the sort of image of science which is open to sceptical attack from a basis which is rooted in the metaphysics of the inappropriate conception of knowledge. For given that conception, and given the nature of contemporary science, objective knowledge must be revealed as unattainable by scientific means. This would seem to be Feyerabend's line of argument. From his standpoint perpetuation of the image of science as yielding objective knowledge is ideological, and is, like science itself, merely a perpetuation of values which have their origin outside science.

Only if it is admitted that part of what constitutes cognitive progress is rectification of the conception of objective knowledge itself, of standards of objectivity, in which the boundary between subjective and objective is redrawn, can contemporary science, without false consciousness, continue to see itself as having made, and as capable of continuing to make, cognitive progress. And the price of this self-legitimation is the discarding of dogmatism: recognition internal to the image of science (within the account of objective knowledge) of the possibility of further improvement, of future revision in the light of the unending dialogue between theory and experiment, which is driven by the rational requirements of unity and consistency, imposed by a community of scientists reflecting critically upon the experimental and theoretical practices in which they participate.

6 FROM SCIENCE TO THE PHILOSOPHY OF SCIENCE

We are now, finally, in a position to see the full force of the claim that philosophy of science should take its cues from science and not seek to impose prior philosophical theories of epistemology and metaphysics on science. Bachelard argues that science is still natural philosophy: 'a philosophical activity accompanies scientific activity, today as at the time of Leibniz' (ARPC p.268). 'But the sense of the philosophical evolution of scientific ideas is so clear that one must conclude that scientific knowledge orders thought, that science orders philosophy itself. Scientific thought thus furnishes a principle for the classification of philosophies and for the study of the progress of reason' (PN p.22).[16] But all that has been said in this chapter to bring out the force of this claim, and to show how it arises from and fits in with the aspects of Bachelard's position discussed in previous chapters, has been predicated upon the assumption that the form of causal laws *has* changed. For it is this which plays the crucial role in arguments to the effect that twentieth-century changes in science, and in physics in particular, have involved changes in epistemology and metaphysics,

[16] This attitude is echoed in Einstein and Infeld 1961 p.51.

changes in the conception of objective knowledge and of methods of attaining it, and in the conception of the nature of physical reality. This assumption in turn rests on the claim that mathematics provides the rational framework for contemporary scientific thought; that the forms of causal laws are mathematical as opposed to logical forms. If the involvement of mathematics in contemporary science were merely a matter of the use of logically formalised and axiomatised theories, treated as formal calculi (computing mechanisms), then there would be no basis for claiming that the form of causal laws has changed in any fundamental way. The forms would still be ultimately logical forms. Mathematics does not, on such a view, introduce any new forms of thought or any new categories by means of which to think about physical reality.

The acceptance and successful application of non-Euclidean geometries was seen by many as putting a final nail in the coffin of rationalist paradigms of knowledge, by showing that even in geometry there is no possibility of synthetic *a priori* knowledge. This seems to amount to a recognition that there is no *knowledge* in mathematics. Mathematics is thus shown by the course of its own historical development, to have a purely formal role in relation to empirical science; it merely provides a convenient calculus, a set of techniques, but has no real cognitive role. From this point of view it would seem that consideration of mathematics and its epistemology can drop out of the epistemology of contemporary science, for strictly speaking it should now be recognised that there is no epistemology of mathematics; it is an activity of which some other account must be given. Equally, because mathematics has nothing to do with cognitive content, the mathematical form of a physical theory cannot contribute to its factual content; in principle the empirical content should be separable from the mathematical form.

However, as we have seen, Bachelard insists on the inseparability of mathematical form from empirical content. But what we have also found is that in order to fill in details of the way in which abstract, mathematically formed theoretical notions find application in experience we have to presuppose that mathematics does provide the rational frameworks for developed sciences and that the import of this claim has to be spelt out by reference to a view of the nature and role of mathematics that is, although not Kant's, still Kantian in spirit. How can this question of the relation between mathematics and science, crucial to the positions outlined in this chapter and the last, be settled?

It might seem that one could at least resolve the question of the separability or otherwise of mathematical form from empirical content by appeal to examples drawn from science itself. It seems possible to claim that in principle any axiomatised physical theory could be reformulated as a formal first-order theory without the use of singular terms ranging over mathematical entities such as numbers, sets or functions. Putting aside the technical difficulties raised by trying to show whether this is in fact the case for specific theories (for details see Field 1980, also Sneed 1971), one can still ask whether the successful completion of such a task would conclusively demonstrate the separability of mathematical form from empirical content. The problem is that the whole idea that a theory can be reformulated in such a way that its empirical content is not affected already presupposes the separability of form from content. For it presumes that it is possible for two axiomatic formulations, expressed in different formal languages, to be formulations of the same empirical theory. So it must set some standard for judging sameness of (empirical) content across differences of form, thereby portraying content as independent of form. For Bachelard, on the other hand, formalisation or reformalisation is a reflective activity, is itself a move in the epistemological game, one which does not leave concepts (or their extensions) unchanged. To change the form of a theory is at the same time to change its content. We are thus thrust back to the basic difference between logical and epistemological analyses of science (discussed in Chapter 1).

Equally, if it were possible to argue that for a given theory a non-mathematical reformulation is impossible, there would still be a choice of conclusions to be drawn, a choice between

(1) treating the theory in the same way as a mathematical theory, giving an instrumentalist account of its development and use, and (2) admitting some cognitive role for mathematics in science. Which choice is made is likely to depend on the view taken of the nature of mathematics. On any formalist view of mathematics, the former seems to be the only available conclusion. This may be one reason why many philosophers of science have assumed that scientists have taken an instrumentalist attitude towards quantum mechanics, for this is one case where it would seem that there is no way of eliminating the mathematics. On the other hand, Bachelard can insist on the mathematical character of the theory while at the same time opposing instrumentalist accounts of it because he gives mathematics a direct cognitive role in science.

The whole difficulty in settling the issue lies in getting agreement about how cognitive content is to be assessed. Differences of opinion on this can be traced back to the models of knowledge and of scientific theorising presumed, that is to the difference between a theory as a deductively closed set of sentences and a theory as a conceptual framework structuring thought conceived as an activity (i.e. as structuring a rational practice). In other words, it goes back to the difference between passive and active conceptions of thought and knowledge. This is in turn rooted in different views of objectivity. If it is the case that passivity on the part of the subject is a necessary condition of objective knowledge, then emphasis in discussions of knowledge will be on cognitive *states* and on the statements in which these are expressed. Knowledge will be objective to the extent that these states are determined by the objects of knowledge (as in direct, non-distorting perception), or to the extent that the corresponding statement *is* determinately true or false in virtue of the state of that to which its referential terms refer. But Bachelard argues that contemporary science has turned its back on this conception of objectivity, and that it finds objectivity through the subject's activity. This is why there is an emphasis on mathematics, for mathematics, as characterised by Bachelard, deals with the forms of active, constructive, but also paradigmatically rational thought. The emphasis throughout is on method, on activity, whether at the theoretical or at the experimental level. Objective scientific knowledge is not a product of confrontation with a given object; the object is not the locus of objectivity. Objective knowledge is the product of scientific activity, the activity which generates a sequence of successively rectified concepts which there is reason to believe are convergent, and hence is one which can be taken as a sequence of closer approximations to fully objective knowledge. The sequence is the source of the conception of the object (as its limit) and of the judgement that it is a sequence in which progress has been made, that it has a limit (is convergent), that this limit is better known, more accurately understood, than it was, but is something about which there is more, and more accurate, knowledge yet to be gained. Such a judgement is only possible on the basis of a (possibly implicit) grasp of the activities which generate the sequence and of the nature of the empirical and rational constraints (the standards of objectivity) these impose on the addition of successive terms.

Yet it is important to understand the level at which there is disagreement here. For what has created something of a crisis in the understanding of science within analytic (or post-Fregean) philosophy is an increasing appreciation of many of the characteristics of science to which Bachelard points, an increasing realisation of the extent to which scientists are active in their pursuit of knowledge. Against the background of a philosophic framework in which objectivity is linked to the passivity of the subject this leads inevitably to serious doubts about the extent to which there is, or can be, objective knowledge in science.

Bachelard argues that this is the wrong way to approach the matter, that one should derive one's philosophy of science from science itself, rather than seek to impose an antecedent philosophically acquired account of objective knowledge on science. Yet it seems that his view itself arises out of a philosophic tradition, out of an approach to knowledge which had already insisted on the subject's activity as a necessary condition of the possibility of objective knowledge. So in what sense can it be said that his philosophy of science *is* derived from science itself rather than being imposed on it?

Indeed, methodological consistency requires that Bachelard recognise that there is some philosophical input into his position, for if he is to be read as maintaining an internally consistent position, then he cannot be read as claiming that it is possible to approach any science in a philosophically neutral way in order to present an account of its philosophy. This is no more possible than pure observation in science, and indeed a good deal less so. The epistemological exposition of scientific theories is itself an interpretative and reflective enterprise.

The sense in which there is, nevertheless, some justification for the claim to be taking philosophical cues from science is the sense in which (as explained in Chapter 2) non-Cartesian epistemology is pluralistic. It is an abstract epistemological framework in which philosophies of different sciences at different stages of their development can be discussed. (This is analogous to the situation with respect to geometry. Topology provides a framework within which different geometries can be discussed, and within which different characteristics of space, or of space–time, employed by different scientific theories, can be compared. It opens up the possibility of discussing the constraints on the choice of one geometry rather than another. Yet it does so still within a particular kind of more general and abstract framework, one which is not the only conceivable abstract generalisation from Euclidean geometry.) It is one within which Cartesian epistemology and metaphysics can (like other empiricist and realist positions) be characterised, discussed and understood, in that reasons can also be given for its rejection. Cartesian philosophy *is* the philosophy of Cartesian science. That, from Bachelard's point of view, is precisely its problem. Cartesian science is so radically different from contemporary science that a philosophy appropriate to it will not be appropriate for contemporary science.

The argument that philosophy should take its cues from science is thus the argument for the move to non-Cartesian epistemology. It is in the interpretation of the shortcomings of Cartesian epistemology and in the idea that the way to move from a Cartesian position is by dialectical generalisation that prior philosophical commitments are revealed, commitments concerning the nature of the rational subject and of rational objectivity. The idea that philosophies of science can and do change with the development of science, that scientific developments may reflect back onto epistemology and metaphysics, is a consequence of the way in which Bachelard interprets the need to move away from Cartesian epistemology.

The (by now widely recognised) need to reject the Cartesian quest for epistemological foundations is seen by Bachelard as requiring in addition a rejection (1) of the perceptual or quasi-perceptual model of objective knowledge (that which grounds objectivity (a) in objects of perception, and (b) in the subject's passivity) and (2) of the assumption that the intellect (the rational subject) can be completely known and characterised *a priori*. It is this latter which is crucial to the idea that philosophy must change with science, for it amounts to a denial of the idea that epistemology can start from *a priori* knowledge of the knowing subject and so to a denial that it can take as given a clear separation between subjective and objective in empirical knowledge. Instead, the progress of objective knowledge involves repeatedly redrawing the boundaries between subjective and objective; revision in conceptions of objective reality have a necessary impact on the subject and his view of himself. The subject is, in consequence, never absent from Bachelard's epistemology, even though the quest for objective knowledge involves successive attempts at transcendence of the subjectively conditioned viewpoint. It is projection of a subjective–objective polarity which gives rise to the space within which progress can take place. To eliminate the subject from epistemology (which is the epistemological analogue of removing the subject from the world of which he seeks knowledge) is to annihilate this space.

Post-Fregean philosophy has turned away from Cartesian philosophy in a different direction: by rejecting the quest for epistemological foundations and the procedures of epistemological analysis and replacing them by the quest for logical foundations and the procedures of logical analysis. (Analysis in the context of justification (see Chapter 1) is still foundational in intent, revealing what are the logical foundations of a given theory or discipline, but because the analysis is not epistemological such analyses say nothing about the cognitive status of what is logically basic.) A principle concern has been avoidance of 'psychologism'. The move to analysis of language as the means to the analysis of (rational, objective) thought (see p.5) was dictated by this concern. But in making this move it is presumed that the structures and functions of language can (as Cartesian intellectual thought before it) be fully known without reference to the world. Logic and philosophy of language are non-empirical disciplines which have displaced Cartesian epistemology, but which perform a similar function of providing the framework within which the problem of the nature of objective empirical knowledge must be posed and discussed.

The initial separation of subjective from objective is made in the sharp separation of logic from epistemology, of logical from epistemological issues. The separation is thus not questioned, but reinforced by eliminating, from the start, all reference to the subject. Language, as studied, is the appropriate vehicle for objective knowledge because the subjective components in the understanding of language have been eliminated from the concept of meaning employed. The advance from an objective thought (the cognitive content of a sentence) to objective knowledge is the determination of the truth value to be attached to that thought in the light of how things are in the world. From the perspective of linguistically formulated thought, progress in the acquisition of objective knowledge appears as a matter of determining the truth values of sentences. There is thus a new kind of epistemology which comes after the work of logical and/or conceptual analysis, an epistemology centred on the problem of the mechanisms of truth value determination, a wholly objective epistemology, one from which the knowing subject has disappeared. Thus the norms set by the direction of Frege's rejection of psychologism not only build in a presumed initial separation of subjective from objective, but also a conception of what objective knowledge *can* consist in.

From the standpoint of post-Fregean philosophy, Bachelard's account of scientific objectivity can only receive a negative evaluation. His procedures do not conform to its norms, his account of objectivity based in the subject's activity violates its conception of what objective knowledge could consist in. The subjective–objective space within which Bachelard sees science as working and progressing does not exist for (and so cannot be seen from) this standpoint. Bachelard's responses to the charge of psychologism and to the claim that critical reflection cannot be both a rational process and a source of innovation have been seen to rest on conceptions of rationality, of the rational subject and his nature, and of the role of mathematics which cannot be incorporated into this framework.

But equally we have seen that from Bachelard's standpoint there are good reasons why the Fregean turn should be negatively evaluated. Reasons for thinking that there are ways other than Frege's for avoiding the charge of an objectionable psychologism (one which would render the acquisition of objective knowledge impossible). Accounts of objectivity and rationality are as tightly linked for Bachelard as for Frege; where Frege links reason to logic, to deduction and the establishment of objective truths, Bachelard links reason to planned, justifiable activity and objectivity to recognition of the nature of the constraints on and obstacles to successful action, recognition made possible by critically reflective evaluation of both plans and actions. From Bachelard's standpoint, it is epistemology without a knowing subject that is an impossibility.

Thus the fact that there is a philosophic input shaping Bachelard's conception of the task of philosophy of science and thus the kind of account which he offers of contemporary science, and the fact that this is at odds with that which has shaped the analytic philosopher's conception of the task raises difficult problems of evaluation. These are not just different accounts of science, but different positions on how even to approach science philosophically, where each position contains within it grounds for rejecting the other.

REFERENCES

Althusser, L. 1977. *For Marx*, trans. B. R. Brewster, London: New Left Books (first French edn 1966).

Anscombe, G. E. M. 1976. 'The Question of Linguistic Idealism', in 'Essays on Wittgenstein in honour of G. H. von Wright', *Acta Philosophica Fennica*, 28.

Ayer, A. J. 1973. *The Central Questions of Philosophy*, London: Weidenfeld & Nicolson.

Bacon, F. 1960. *The New Organon and Related Writings*, ed. F. H. Anderson, Indianapolis: Bobbs-Merrill (reprint of the J. Spedding and R. L. Ellis translation, Boston: 1863, first published as *Novum Organon*, London, 1620).

Bhaskar, R. 1975. 'Feyerabend and Bachelard: Two Philosophies of Science', *New Left Review*, 94.

Bloor, D. 1976. *Knowledge and Social Imagery*, London: Routledge & Kegan Paul.

Bohm, D. 1957. *Causality and Chance in Modern Physics*, London: Routledge & Kegan Paul.

Bohm, D. 1980. *Wholeness and the Implicate Order*, London: Routledge & Kegan Paul.

Bohr, N. 1949. 'Discussion with Einstein on Epistemological Problems in Atomic Physics', in *Albert Einstein: Philosopher–Scientist*, I. ed. P. A. Schilpp, La Salle, Ill.: Open Court.

Boyle, R. 1911. *The Sceptical Chymist*, London: Dent (first edn 1661).

Boyle, R. 1979. 'The Origin of Forms and Qualities According to the Corpuscular Philosophy (publ. 1666)', in *Selected Philosophical Papers of Robert Boyle*, ed. M. A. Stewart, Manchester: Manchester University Press.

Brouwer, L. E. J. 1964a. 'Consciousness, Philosophy and Mathematics' (1940) in *Philosophy of Mathematics: selected readings*, ed. P. Benacerraf and H. Putnam, Englewood Cliffs, New Jersey: Prentice-Hall.

Brouwer, L. E. J. 1964b. 'Intuitionism and Formalism' (1912) in *Philosophy of Mathematics: selected readings*, ed. P. Benacerraf and H. Putnam, Englewood Cliffs, New Jersey: Prentice-Hall.
Cassirer, E. 1923. *Substance and Function & Einstein's Theory of Relativity*, Chicago: Open Court.
Cohen, P. J. 1966. *Set Theory and the Continuum Hypothesis*, New York: W. A. Benjamin.
Crosland, M. P. 1962. *Historical Studies in the Language of Chemistry*, London: Constable.
Dauben, J. W. 1979. *Georg Cantor: His Mathematics and Philosophy of the Infinite*, Cambridge, Mass.: Harvard University Press.
Davidson, D. 1973. 'On the Very Idea of a Conceptual Scheme', *Proceedings and Addresses of the American Philosophical Association*, 47.
Davis, H. M. 1952. *The Chemical Elements*, Washington, D.C.: Science Service Inc.
Descartes, R. 1931. *Philosophical Works of Descartes*, I, trans. and ed. E. S. Haldane and G. R. T. Ross, Cambridge: Cambridge University Press.
Descartes, R. 1954. *Descartes: Philosophical Writings*, trans. and ed. G. E. M. Ansombe and P. T. Geach, London: Nelson.
Descartes, R. 1965. *Discourse on Method, Optics, Geometry, and Meteorology*, trans. P. J. Olscamp, Indianapolis: Bobbs-Merrill.
Duhem, P. 1962. *The Aim and Structure of Physical Theory*, trans. P. P. Weiner (first published as *La Théorie Physique: Son Objet, Sa Structure*, Paris: 1914), New York: Atheneum.
Dummett, M. 1973a. 'The Justification of Deduction', *Proceedings of the British Academy*, LIX.
Dummett, M. 1973b. *Frege: Philosophy of Language*, London: Duckworth.
Dummett, M. 1978. 'Can Analytic Philosophy be Systematic, and Ought it to Be?' in *Truth and Other Enigmas*, London: Duckworth.
Einstein, A. 1920. *Relativity: The Special and the General Theory*, London: Methuen.
Einstein, A. & Infeld, L. 1961. *The Evolution of Physics*, Cambridge: Cambridge University Press. First edn, Cambridge 1938.
Feigl, H. 1970. 'The "Orthodox" View of Theories', in *Minnesota Studies in the Philosophy of Science, IV: Analysis of Theories and Methods of Physics and Psychology*, ed. M. Radner and S. Winokur, Minneapolis: University of Minnesota Press.
Feyerabend, P. K. 1965. 'On the "Meaning" of Scientific Terms',

Journal of Philosophy, 12, pp.266–74.

Feyerabend, P. K. 1975. *Against Method*, London: New Left Books.

Feyerabend, P. K. 1981. *Problems of Empiricism: Philosophical Papers*, 2, Cambridge: Cambridge University Press.

Field, H. 1980. *Science Without Numbers*, Oxford: Blackwell.

Fleck, L. 1979. *Genesis and Development of a Scientific Fact*, ed. T. J. Trenn and R. K. Merton, trans. F. Bradley and T. J. Trenn, Chicago: University of Chicago Press (originally published as *Entstehung und Entwicklung einer wissenschaftlichen Tatsache: Einführung in die Lehre vom Denkstil und Denkkollectiv*, Switzerland, 1935).

Fowler, D. H. 1979. 'Ratio in Early Greek Mathematics', *Bull. Am. Math. Soc. (new series)* 1, No. 6, pp.807–46.

Fraenkel, A. A. 1922. 'Der Begriff "definit" und die Unahängigkeit des Auswahlsaxioms', *Sitzungsberichte der Preussischen Akademie der Wissenschaften, Physicalisch-Mathematikertische Klasse 1922* (pp.253–7). English translation 'The notion "definite" and the independence of the Axiom of Choice', in J. van Heijenhoort (ed.) *From Frege to Gödel*, Cambridge Mass.: Harvard University Press, 1967, pp. 284–9.

Frege, G. 1959. *The Foundations of Arithmetic*, trans. J. L. Austin, Oxford: Blackwell. First published as *Die Grundlagen der Arithmetik*, Breslau, 1884.

Frege, G. 1960. *Translations from the Philosophical Writings of Gottlob Frege*, ed. P. Geach and M. Black, Oxford: Blackwell.

Frege, G. 1971. *On the Foundations of Geometry and Formal Theories of Arithmetic*, trans. E. W. Kluge, New Haven Conn.: Yale University Press.

Frege, G. 1979. *Posthumous Writings*, ed. H. Hermes, F. Kambartel, F. Kaulbach, trans. P. Long and R. White, Oxford: Blackwell.

Galileo, G. 1638. *Discorsi e Dimonstrazion: Matematiche, intorno à due nuoue Scienze*, Leyden. Modern edition *Dialogues Concerning Two New Sciences*, trans. H. Crew and A. de Salvio, New York: Dover 1954.

Gaukroger, S. W. 1976. 'Bachelard and the Problem of Epistemological Analysis', *Studies in History and Philosophy of Science*, 7, pp.189–244.

Gombrich, E. H. 1963. *Meditations on a Hobby Horse*, London: Phaidon.

Goody, J. 1977. *The Domestication of the Savage Mind*, Cambridge: Cambridge University Press.

Hacking, I. 1982. 'Language, Truth and Reason', in M. Hollis and S.

Lukes (eds) *Rationality and Relativism*, Oxford: Blackwell.

Hardy, G. H. 1952. *A Course of Pure Mathematics*, Cambridge: Cambridge University Press.

Hempel, C. G. 1965. *Aspects of Scientific Explanation*, New York: The Free Press.

Hertz, H. 1956. *The Principles of Mechanics Presented in a New Form*, trans. D. E. Jones and J. T. Walley, New York: Dover. Original German edition as vol. 3 of Hertz's collected works publ. 1894.

Hesse, M. 1966. *Models and Analogies in Science*, Notre Dame: University of Notre Dame Press.

Hesse, M. 1974. *The Structure of Scientific Inference*, London: Macmillan.

Jung, C. G. 1968. *Psychology and Alchemy*, London: Routledge & Kegan Paul. Originally published in German as *Psychologie und Alchemie*, Zurich, 1944.

Kant, I. 1929. *Critique of Pure Reason*, trans. N. Kemp Smith, London: Macmillan. First edition *Kritik der reinen Vernunft*, Riga, 1787.

Kant, I. 1970. *Metaphysical Foundations of Natural Science*, trans. J. Ellington, Indianapolis: Bobbs-Merrill. Originally published as *Die metaphysischen Anfangsgründe der Naturwissenschaft*, 1786.

Kripke, S. 1972. 'Naming and Necessity', in *Semantics of Natural Language*, ed. G. Harman and D. Davison, Dordrecht: Reidel.

Kuhn, T. 1962. *The Structure of Scientific Revolutions*, Chicago: University of Chicago Press.

Kuhn, T. 1964. 'A Function for Thought Experiments', in *L'Aventure de la science, Mélanges Alexandre Koyré*, Paris: Hermann. Reprinted in Kuhn 1977.

Kuhn, T. 1970. 'Reflections on my Critics', in *Criticism and the Growth of Knowledge*, ed. I. Lakatos and A. Musgrave, Cambridge: Cambridge University Press.

Kuhn, T. 1977. *The Essential Tension*, Chicago: University of Chicago Press.

Lakatos, I. 1976. *Proofs and Refutations*, Cambridge: Cambridge University Press.

Lakatos, I. 1978. *Mathematics, Science and Epistemology: Philosophical Papers*, 2, ed. J. Worrall and G. Currie, Cambridge: Cambridge University Press.

Laudan, L. 1977. *Progress and its Problems*, Berkeley: University of California Press.

Lecourt, D. 1975. *Marxism and Epistemology*, London: New Left Books.

Mach, E. 1960. *The Science of Mechanics*, trans. T. McCormack, La Salle, Illinois: Open Court. First German edition *Die Mechanik in Ihrer Entwicklung Historisch-Kritisch Dargestellt*, 1883.

Mackie, J. L. 1977. *Ethics: Inventing Right and Wrong*, Harmondsworth: Penguin.

Mill, J. S. 1971. *Autobiography*, ed. J. Stillinger, Oxford: Clarendon Press. First published 1873.

Mill, J. S. 1973. *A System of Logic* in *The Collected Works of John Stuart Mill*, VII, Toronto: Toronto University Press, and London: Routledge & Kegan Paul. First published 1843.

Nagel, E. 1961. *The Structure of Science*, London: Routledge & Kegan Paul.

Newton-Smith, W. 1981. *The Rationality of Science*, London: Routledge & Kegan Paul.

Piaget, J. 1966. with E. W. Beth, *Mathematical Epistemology and Psychology*, trans. W. Mays, Dordrecht: Reidel.

Popper, K. R. 1959. *The Logic of Scientific Discovery*, London: Hutchinson.

Popper, K. R. 1963. *Conjectures and Refutations*, London: Routledge & Kegan Paul.

Popper, K. R. 1972. *Objective Knowledge: An Evolutionary Approach*, Oxford: Clarendon Press.

Popper, K. R. 1983. *Realism and the Aim of Science*, from the *Postcript to the Logic of Scientific Discovery*, ed. W. W. Bartley III, London: Hutchinson.

Putnam, H. 1978. 'Reference and Understanding', in his *Meaning and the Moral Sciences*, London: Routledge & Kegan Paul.

Quine, W. V. 1963. 'Two Dogmas of Empiricism', in his *From a Logical Point of View*, New York: Harper & Row.

Reichenbach, H. 1938. *Experience and Prediction*, Chicago: Chicago University Press.

Rorty, R. 1980. *Philosophy and the Mirror of Nature*, Princeton: Princeton University Press.

Russell, B. 1897, *Foundations of Geometry*, Cambridge: Cambridge University Press.

Russell, B. 1917. *Mysticism and Logic*, London: Allen & Unwin.

Russell, B. 1919. *Introduction to Mathematical Philosophy*, London: Allen & Unwin.

Russell, B. 1937. *The Principles of Mathematics* (2nd edn), London: Allen & Unwin (first edn 1903).

Russell, B. 1972. 'The Philosophy of Logical Atomism' (first published 1918) in *Russell's Logical Atomism*, ed. D. F. Pears, London: Collins.

Shortley, G. & Williams, D. 1965. *Elements of Physics*, Englewood Cliffs, New Jersey: Prentice-Hall.

Sneed, J. D. 1971. *The Logical Structure of Mathematical Physics*, Dordrecht: Reidel.

van Fraassen, B. C. 1980. *The Scientific Image*, Oxford: Clarendon Press.

Whitehead, A. N. 1925. *Science and the Modern World: Lowell Lectures, 1925*, New York: Macmillan.

Wittgenstein, L. 1963. *Philosophical Investigations*, trans. G. E. M. Anscombe, Oxford: Blackwell.

Wittgenstein, L. 1967. *Remarks on the Foundations of Mathematics*, ed. G. H. von Wright, R. Rhees, G. E. M. Anscombe, trans. G. E. M. Anscombe, Oxford: Blackwell.

Zermelo, E. 1908. 'Untersuchungen über die Grundlagen der Mengenlehre I' *Mathematische Annalen*, 65. English translation 'Investigation into the foundations of set theory I', in *From Frege to Gödel*, ed. J. van Heijenhoort, Cambridge, Mass.: Harvard University Press, 1967.

APPENDIX: BIOGRAPHICAL NOTE

Gaston Bachelard had an unusual career, and perhaps because of this and his own forceful personality, his work has a very independent character. In style and in range of subject-matter it is academically unconventional.

Bachelard was born on 27 June 1884 in Bar-sur-Aube, in Champagne. Here he spent his childhood. From 1903 until the outbreak of the first world war he worked for the postal service, whilst also pursuing scientific studies. He received his *licence* in mathematics in 1912 and in 1913 had obtained a scholarship in mathematics from the Lycée Saint-Louis. But in August 1914, less than a month after marrying a school teacher from his home region, he was mobilised. For his service in the trenches he received the Croix de Guerre.

In 1919 he returned to Bar-sur-Aube to take up a teaching post at the local Collège, where he taught physics and chemistry. In the meantime he had become interested in philosophy and acquired his *licence* in philosophy in 1920, the same year in which his wife died, leaving him with a daughter, Suzanne. He remained in Bar-sur-Aube, continuing both his teaching and his studies in philosophy, obtaining his *agrégation* in 1922, and his doctorate from the Sorbonne in 1927. His doctoral dissertations (*Essai sur la connaissance approchée* and *Etude sur l'evolution d'un problème de physique: la propagation thermique dans les solides*) were directed by Abel Rey and Léon Brunschvicg.

After obtaining his doctorate he began to teach two courses in philosophy every fortnight at the University of Dijon, where he was appointed to the chair of philosophy in 1930. Here he taught for ten years until he was appointed to succeed Abel Rey to the chair of history and philosophy of science at the Sorbonne. He was appointed on the basis of the reputation he had established through his critique of scientific knowledge. And yet shortly after taking up his post at the Sorbonne Bachelard began to publish works of a quite different character (*L'Eau et les rêves* (1942), *L'Air et les songes* (1943), *La Terre et les rêveries de la volonté* and *La Terre et les rêveries du repos* (1948)) in which he turned from reason and science to imagination and poetry, much to the amazement, not to say disquiet, of his academic colleagues. He did, however, continue to write both on the philosophy of science and on poetry. Even after his retirement from the Sorbonne in 1954 he continued to publish and to lecture on a part-time basis. In 1955 he was elected to the *Académie des Sciences morales et politiques* and in 1961 he received the *Grand Prix National des Lettres*. Bachelard died in Paris on 16 October 1962 and was buried in Bar-sur-Aube. He is remembered for his remarkable intellect, his personal charm, for his approachability and his attentiveness to his students, and for his curious blend of down-to-earth practicality and flights of speculative, imaginative thought.

INDEX